Most Wonderful in the Smallest

MAXIME MIRANDA IN MINIMIS – H. C. Richter

Frontispiece. This wonderful watercolor drawing of thirty-five microscopic organisms titled *Maxime Miranda in Minimus* (Most Wonderful in the Smallest) was rendered in 1869 by Henry Constantine Richter (1821–1902), a well-known British zoological illustrator. It is reproduced here from the original by courtesy of the Quekett Microscopical Club, J.A. Dutton, Curator, FRMS, MRI.

Most Wonderful in the Smallest

A Year in Pursuit of
Common Freshwater Microorganisms

by
Linda VanAller Hernick

The McDonald & Woodward Publishing Company
Newark, Ohio

Most Wonderful in the Smallest

www.mwpubco.com

Most Wonderful in the Smallest
A Year in Pursuit of Common Freshwater Microorganisms

All rights reserved; first printing May 2017
This book has been printed in the United States of America by
McNaughton & Gunn, Inc., Saline, Michigan, on paper that
(1) meets the minimum requirements of permanence for printed
library materials and (2) is free of harmful chemical contaminants.

10 9 8 7 6 5 4
27 26 25 24 23 22 21

Library of Congress Cataloging-in-Publication Data

Names: Hernick, Linda VanAller, author.
Title: Most wonderful in the smallest : a year in pursuit of common
freshwater microorganisms / by Linda VanAller Hernick.
Description: Newark, Ohio : The McDonald & Woodward Publishing
Company, [2017] | Includes bibliographical references and index.
Identifiers: LCCN 2017018850 | ISBN 9781935778370 (pbk. : alk. paper)
Subjects: LCSH: Freshwater microbiology—New York (State)
Classification: LCC QR105.5 .H47 2017 | DDC 579/.17747—dc23
LC record available at https://lccn.loc.gov/2017018850

Contents

Dedication

For John

and in memory of my parents

Preface

Summers in the Schoharie Valley of New York state where I grew up were the best part of childhood. Day flowed into day in a seemingly endless adventure of reckless bike-riding, treehouse-building, and wandering. But the humid days of August slowed one down and I spent them quietly, leaning against a round hassock in the cool living room, reading. It was at this point in the summer when, through some nearly divine plan, my annual package of *Scholastic Book Club* books would arrive. These had been painstakingly selected and paid for with saved allowance money in June before school closed. Even though I had kept the little four-page catalogue with my choices circled in anticipation, delivery of the books was a marvelous surprise. To see the real colors of the covers in contrast to the black and white catalogue images and to know that the books were mine was pure bliss.

One summer, in addition to *Mystery in the Pirate Oak, The Peculiar Miss Pickett,* and *Emil and the Detectives,* I received *Pioneer Germ Fighters* (first published in Britain in 1962 as *Pioneers Against Germs*) by Navin Sullivan. In the selection of this book my attention had immediately been drawn to its cover which featured, in the center, a microscope. The pen and ink drawing by distinguished British illustrator Eric Fraser detailed microbiologist Alexander Fleming in his 1928 London laboratory, inoculating and examining the results of agar plates. This pursuit would ultimately lead to his ground-breaking discovery of penicillin. The superb quality of my 35¢ children's paperback succeeded in transmitting one of its messages home — the microscope is an amazing instrument!

As a consequence of acquiring Navin Sullivan's book as a child, microscopy suffused my entire adult working-life. From screening cell samples for the presence of cancer, to scrutinizing disaggregated rock samples for Cambrian-age microfossils, much of my view of the world for nearly four decades was through the lens of a microscope. Now, retired, my interest has become a passion for light microscopy and the world of freshwater microorganisms. However, my means of study are simple. While scanning probe microscopy on the nanoscale may be the latest technology, there is still a place for an individual (like myself) with a basic light microscope. And even though "small world" photography competitions are now mostly dependent on sophisticated techniques such as "differential interference contrast with image stacking" it is possible for the non-professional to obtain photos too. Some small "point and shoot" digital cameras placed directly over a microscope's eyepiece can capture images that are sufficient to delight oneself and be instructive. Certain Smartphones also lend themselves to this type of quick image capture. For the truly dedicated, a trinocular microscope with a DSLR camera, T-mount adapter, and relay optics can produce good-quality micrographs.

Most Wonderful in the Smallest is a year-long observation of single-cell organisms and tiny invertebrate animals that thrive in freshwater environments near my home and that are common all over North America. It is written in the form of a journal comprising four chapters — "Spring," "Summer," "Autumn," and "Winter." Journal entries describe and illustrate the forays I make for collecting water samples, and what I find of interest in the samples themselves. Most organisms present in the samples range in size from a "hair's breadth," or roughly three one-thousandths of an inch, to one or two millimeters. However, for the sake of convenience and accuracy, actual size measurements under one millimeter are given in micrometers (a micrometer is equal to one millionth of a meter) denoted by the symbol "µm."

This book is an invitation to anyone who has ever contemplated a pond, ditch, stream, or swamp and wondered what lies there beyond our limited vision. It beckons those who for some unknown reason are drawn to microscopes. It is a call to *everyone* to see the amazingly beautiful and diverse tiny organisms that dwell in all water bodies but for whom no specific conservation measures are ever taken. Wetlands of all types remain in constant peril and the risk of losing these nurseries for amphibians, reptiles, and other macro-species has been well-documented. However, just as importantly, in the destruction of wetlands and pollution of waters we risk also the loss of whole and balanced *micro-communities*.

It is my hope that **Most Wonderful in the Smallest** will begin to acquaint readers with a portion of Earth's vast microscopic biodiversity. Acquaintance and understanding lead to appreciation. Appreciation engenders protection.

About the Micrographs

One of my purposes for this book is to demonstrate that photographs of microlife can be obtained by simple means. Many of the micrographs, particularly the dark field images, were taken with a Canon *PowerShot A590* camera positioned directly over the eyepiece of an AO Series 10 microscope. Additional micrographs were taken using a Canon *EOS T6s* DSLR camera mounted on a Meiji trinocular microscope. Use of the former photographic technique in combination with live specimens, often on the move, has caused some of the images to reflect imperfections.

The diameter or length of many specimens in the micrographs, and the magnifications used in taking the micrographs, is provided in the captions. As the Canon *PowerShot A590* camera has an optical zoom of 4x, this magnification adds to that indicated in some of the captions.

Biological Terms

The use of some biological terms in this book has been unavoidable and some readers might encounter words whose meanings they will not understand. A glossary is provided at the end of the book to assist with technical terms that appear multiple times in the text. Other terms can be found easily on the internet, and further reading about organisms introduced in this book is encouraged by my inclusion of a bibliography.

Acknowledgements

I would like to express thanks to the following for assistance with the identification of organisms: Dr. Richard Triemer, Dr. Paul Bartels, Dr. James H. Thorp, and Dr.

Preface

Julian Smith III. Any errors in the identification or interpretation of organisms are my own. Thanks are also given to Tony Dutton and Joan Bingley of the Quekett Microscopical Club for their assistance with my obtaining and using the Richter drawing that appears on the front cover and as the frontispiece of this book.

In addition, I must express my appreciation to my husband, John, for his unwavering support in all my endeavors.

Most Wonderful in the Smallest

A Year in Pursuit of
Common Freshwater Microorganisms

Spring

Sunday, March 31ˢᵗ, 5:10 P.M.

An early Easter Sunday. The past season was an old fashioned winter with a good deal of snow and bitter cold. The landscape has now been relieved of remaining snow with current milder temperatures but it is brown, ravaged, and totally spent. The only sign of the new vernal equinox is some green slime emerging in the ditch along my road (Figure 1.1).

Figure 1.1. One of last fall's leaves provides scale to show *Spirogyra* sp. filaments emerging in the ditch. The bright green color is often characteristic of *Spirogyra*.

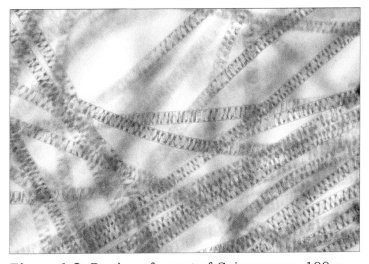

Figure 1.2. Portion of a mat of *Spirogyra* sp. 100x

I visit the ditch with a collecting jar and awkwardly scoop up a sample of slime. The green component is probably *Spirogyra*, one of the earliest filamentous algae to show itself here in spring. No doubt it has been present in the water for some time. However, the warmer sun has induced an increase in photosynthesis as evidenced by the little bubbles of oxygen floating filaments slowly to the surface. The slime is actually mucilage. This is a gelatinous substance made up mostly of carbohydrate polymers and extruded through pores in the alga's cell walls. Mucilage production is common in algae and performs a variety of functions. In the case of *Spirogyra*, it creates enveloping sheaths around the filaments that bind them into floating mats for more exposure to the sun (Figure 1.2). It also aids in water retention, thus preventing desiccation.

I bring the jar into my "lab" which is my kitchen with its 1970s-vintage sink and formica countertop. The end of the countertop juts out just enough to accommodate a stool beneath it and so affords a perfect place for a microscope. My microscope is an American Optical Spencer Series 10, also of 1970s origin, rather timeworn but still usable. I am proud that it bears the name "Spencer" on its small metal nameplate as Charles Achilles Spencer was America's first microscope maker. Until he set up shop in Canastota, New York in 1838, the only quality microscopes available in America were manufactured in England and France. The Spencer company produced instruments with such fine optics that it was awarded a gold medal at the Paris Exhibition in the 1870s.

Spencer Lens Company, managed by Charles Spencer's son, was purchased by the American Optical Company in 1935. AO produced quality microscopes for a number of years. However, by means of a series of business transfers it was merged with other companies.

My AO scope is battleship gray and seemingly bulletproof in its substantial weight and sturdiness. It has three objective lenses — 4x, 10x, and 20x. In addition to standard bright field illumination, the condenser is equipped to allow only oblique light to strike the subject making it bright against a black background. This type of illumination is known as dark field. I often prefer dark field for taking photographs as translucent subjects stand out more readily.

From my sample jar I first read and record the temperature (°F) with my 5½-inch pocket thermometer. With test paper and color chart, I determine and record pH. Today the values are 46° and 7, respectively. Then, with a 7 ml transfer pipette I withdraw a bit of water and sediment from the jar. I place it in a "concavity slide" and, tipping the slide slightly toward me so the water meets the edge of the angled 22 x 22 mm cover glass, drop the glass over the concavity. Concavity slides hold a surprising amount of sample. They also keep organisms happy, that is, they are not squashed, distorted, or mangled by the pressure of a cover glass against an ordinary slide. I place my slide on the microscope stage and take a look using the 10x objective. As suspected, the green slimy stuff is *Spirogyra*. It is magnificent! The ribbon-like chloroplast winds beautifully through each cell (Figure 1.3), the cells oriented end-to-end to form long, unbranched filaments. As I continue to gaze, I am startled to see small nubs that have formed along one edge of some of the filaments, and then I see the reason why. Two filaments have aligned parallel to one another.

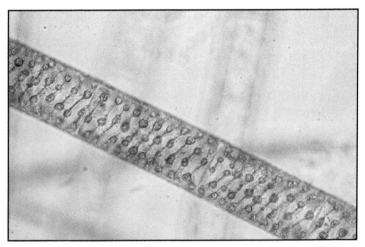

Figure 1.3. *Spirogyra* sp. filament showing the winding chloroplast. The round structures are pyrenoids. 200x

The nubs or conjugation tubes on the filaments have joined and fused to become conjugation canals through which the protoplasm of male cells has transferred to join with that of female cells. The result is the formation of zygospores (figures 1.4 and 1.5). These are round or oval bodies with thick, resistant cell walls. Eventually they will be released from their parent cells and sink to the bottom of the water. At a suitable time they will germinate to form brand new filaments. This form of sexual reproduction is characteristic of Zygnematalean filamentous green algae like *Spirogyra, Mougeotia,* and *Zygnema.*

My luck in being able to witness this reproduction phenomenon is suddenly interrupted by a monstrous thrashing animal that entirely fills the field of view. It is an ostracod wildly flailing its appendages. This has the

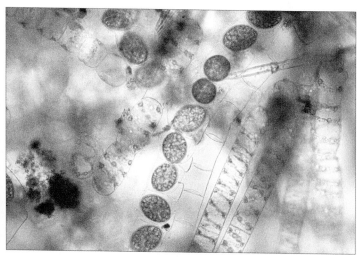

Figure 1.4. *Spirogyra* sp. in conjugation and resulting zygospores. 200x

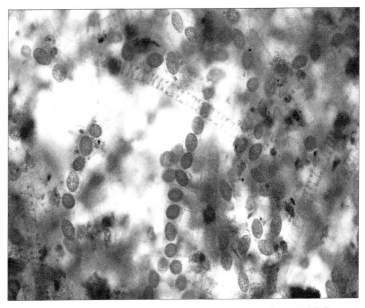

Figure 1.5. *Spirogyra* sp. zygospores galore! 200x

effect of obliterating the *Spirogyra* filaments that were so perfectly disposed. I turn the microscope's turret until the 4x objective is in place. Now the ostracod is not as huge but the chaos from its flailing is more apparent. Filaments are strewn about like storm damage. The ostracod scurries away, it is out of view as quickly as it arrived. However, all is not peaceful yet. Another individual, handsome with brown spots, now takes its place (Figure 1.6). I quickly put the lens of my little Canon *PowerShot* camera over one eyepiece of the microscope. The flash is disabled and I zoom in a little, the zoom preventing the dark circle or "vignetting" around the field of view. I hold the camera steady and wait like a cat — CLICK! While the ostracod relaxes for an instant I get an image.

Ostracods are microcrustaceans inhabiting both fresh and marine waters. North American freshwater species are usually less than 4 mm long. The actual body of these little animals is covered with a carapace consisting of a pair of hinged shells or valves. These open by means of an adductor muscle to allow locomotor appendages to protrude and move. Often the valves are smooth but they may also be pitted or ornamented, or the margins bristled with hairs.

Most ostracods are bottom feeders but some can be found grazing along aquatic plant stems. Their diet consists primarily of algae, bacteria, fungi, and detritus but, in fact, they aren't fussy, their tastes extending all the way to the remains of small dead animals.

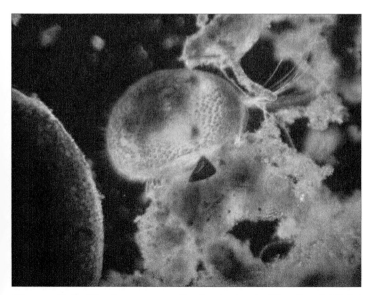

Figure 1.6. The brown-spotted ostracod. 40x, Dark Field

In a given population one can expect to see an overwhelming majority of female ostracods as males are either rare or simply do not exist. This is because many species reproduce solely by means of parthenogenesis, an asexual regenerative mode in which the egg is unfertilized. The individuals present in today's early-season sample may have survived the winter burrowed into the moist ditch sediment in a state of torpor, once again invigorated by warming temperature. Another ostracod survival strategy is the production of desiccation-resistant eggs.

Some ostracod species are unfortunately very sensitive to human-effected assaults on the environment. Changes in carapace shape and chemistry may result from exposure to heavy metals, pesticides, herbicides, and oil spills. Population density and diversity too are subject to alteration from this sort of large-scale chemical carelessness.

Having arisen during the Early Ordovician Period, some 480 million years ago, ostracods litter the fossil record with their carapace remains. Many of these fossils derive from molt stages which accounts for some of their amazing abundance. However, superabundance makes them useful to geologists as index fossils for determining the age of rock units world-wide. Isotopic studies of the valves are also helpful in studies of paleoclimate.

It has been a pleasure to see some examples of these lively little animals in today's sample. The shape of their shells and their many appendages bring to mind their vernacular name, "seed shrimps." The ditch is apparently a hotspot for them and I hope the spring and summer rains are frequent enough to keep the site viable.

I empty my slide into the sample jar and return the jar to the ditch. Gently I pour the water back into its source area marked with pink flagging. While ditch-dwellers like *Spirogyra* and ostracods might seem insignificant, their long life-histories and cosmopolitan distribution make them integral to the planet. They deserve my respect.

Friday, April 10th, 1:15 P.M.

Ostracod Ditch. I am back, curious to see if I can observe more reproduction activity in *Spirogyra*. I scoop up a small mat of filaments while happily noticing a shiny new mass of frog eggs nearby.

Spirogyra does indeed form much of the mat but no conjugation is evident among the filaments. What *is* occurring is reproduction in the yellow-green alga *Vaucheria* that also makes up part of the mat. In contrast to *Spirogyra* filaments which are a series of connected cells separated by cross-walls, *Vaucheria* is characterized by very long, wide, tubular cells called coenocytes (Figure 1.7). A layer of cytoplasm containing many nuclei and disc-shaped plastids lines the inside periphery of the cells; the rest of the space in each cell forms a large vacuole.

The reproductive structures in this species of *Vaucheria*, rather astonishing at first glance, arise from small branches that emerge from the filaments. Several round, green oogonia form a whorl around a clear, curved antheridium on every branch (figures 1.8 and 1.9). Each oogonium contains a single egg which is fertilized by means of sperm swimming from the antheridium and

Figure 1.7. *Vaucheria* sp. coenocytes. 100x

entering a pore in the oogonium wall. The resulting zygote eventually germinates to form another elongate coenocyte filament.

There are more ostracods present today than in my last visit and perhaps a couple of additional species as well. Their frenetic activity in every slide is in extreme contrast to the excruciatingly slow, barely discernible reproduction processes of the algae.

Monday, April 15th, 10:40 A.M.

Time. There is never enough of it. This morning I rush off to the shrub swamp on Rugg Road for a sample (Figure 1.10). I could have walked here in ten minutes but I drive — always the *time.* . .

The shrub swamp is a narrow bit of low, saturated land extending across a portion of a large field. The shrub willows that thrive here produce an incongruous swath of tangled thicket through this otherwise completely open space. There are lots of catkins on the willows today and last season's cattails are completely fluffed out. Redwings flit about and I can hear "peepers" in the distance.

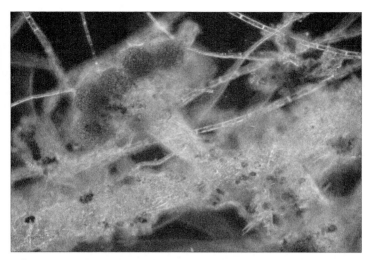

Figure 1.8. Several oogonia on a *Vaucheria* sp. reproductive structure. 100x, Dark Field

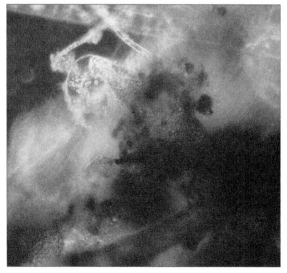

Figure 1.9. The clear, curved antheridium of *Vaucheria* sp. Oogonia are absent. 100x, Dark Field

Figure 1.10. Rugg Road Shrub Swamp.

Though the landscape remains stubbornly bleak, spring has begun to take hold.

It is cloudlessly sunny and I approach the water quietly in the hope of seeing a basking turtle on one of the tussock sedges. No luck. Perhaps it is too early yet. Stepping carefully in among the dry cattail stalks I plunge my collecting jar into the water all the way to the bottom. The water is shallow, perhaps eight inches deep at most, and comes up black with suspended material. I pour several jarfuls into my little plastic aquarium. It is this organic-rich, nutrient-rich water that holds the most promise for micro-diversity.

By the time I have set up my microscope at home the suspended material has settled to the bottom of the aquarium with the clear-water fraction above. However, on the surface I find a small mat of *Riccia fluitans* (Figure 1.11). This is a diminutive thalloid liverwort common in ponds and quiet pools. The plant consists of little translucent green ribbons that divide repeatedly, often interweaving with filamentous green algae. Its species name "fluitans" means floating. Ironically, while *Riccia fluitans* is restricted to aquatic environments, some other species included in this same genus find home in terrestrial habitats such as the soils of fallow farm fields.

Liverworts, like mosses and hornworts, lack food- and water-conducting tissues (phloem and xylem) that are present in vascular plants. However, the simplicity and inconspicuous nature of these plants belie their importance. Undisputed fossils date to the Middle Devonian Period, roughly 380 million years ago, but there is evidence that liverworts may have inhabited the land 100 million years earlier than this making them the oldest extant line of land plants. As such, they very likely

Figure 1.11. The thalloid liverwort *Riccia fluitans*.

hold the additional distinction of being the sister group to all other land plants.

Microscopically, what one sees *most* in freshwater samples are protists. Protists are a large, unwieldy group of eukaryotic organisms that, until recently, comprised Kingdom Protista. (Based on genetic analyses some researchers have now abandoned the traditional "tree of life" and assigned all eukaryotes to eight so-called Supergroups = Archaeplastida, Stramenopila, Alveolata, Rhizaria, Excavata, Amoebozoa, Hacrobia, and Opisthokonta). The Kingdom name refers to their ultra-ancient history as they first appear in the fossil record at least 2 billion years ago. They include all algae, water molds and slime molds, protozoa, and some other ambiguous aquatic organisms. Most protists are unicellular but some, like kelp, are multicellular and can reach lengths of over 100 feet. In addition, certain algal protists form aggregates of cells in the form of colonies, coenobia, or filaments. No matter how they are classified, protists are astounding in the diversity of their morphologies, their reproduction and survival strategies, and, in many cases, their beauty.

What captures my attention in today's first bit of sample is a stunning lunate crescent of *Closterium* (Figure 1.12). This green algal protist is a desmid, the cell comprised of two symmetrical semi-cells separated by a central constriction or "isthmus." The isthmus houses the nucleus. Chloroplasts containing chlorophyll *a* and *b* pigments impart the brilliant green color to the semi-cells but for the tips which enclose vacuoles. Looking closely at the vacuoles with a magnification of 200x I am able to discern energetic movement of the barium

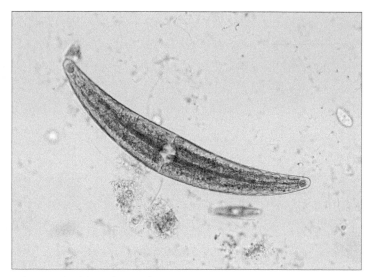

Figure 1.12. *Closterium* sp. ~ 260 µm, 100x

sulfate crystals within (Figure 1.13). That the crystals are present and in constant motion is a phenomenon whose function here is still essentially unknown. Also evident is the constant flow of cytoplasm containing crystals along the entire length of the cell.

The lovely crescent-shaped *Closterium* is the species most frequently seen in my water samples but there are also bow-shaped individuals as well as straight ones looking very needle-like with long spines (Figure 1.14). These species too harbor the cavorting barium sulfate crystals at their tips. Additionally, all of these cells are capable of slow, somersaulting movement, that is, pivoting from their tips in wide arcs. It is believed that mucilage extruded from the ends of the cells facilitates this wonderfully languid means of mobility.

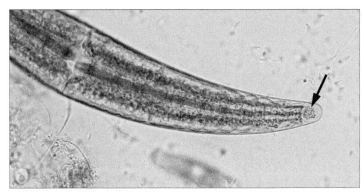

Figure 1.13. *Closterium* sp. showing a vacuole with barium sulfate crystals at the tip. 200x

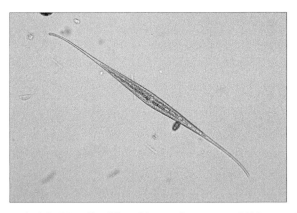

Figure 1.14. Needle-like *Closterium* sp. ~ 250 µm, 100x

Desmids occur as ribbons of cells as well as single cells, for example, *Hyalotheca* (Figure 1.15). These ribbons may sometimes be mistaken for filamentous green algae.

Also conspicuous in today's sample are golden-brown dinoflagellates, a major component of both freshwater and marine plankton. In contrast to the green of

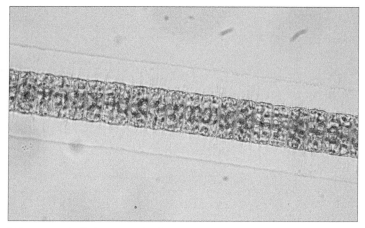

Figure 1.15. *Hyalotheca.* Note the surrounding mucilaginous sheath. 200x

chlorophylls *a* and *b* in *Closterium*, dinoflagellate plastids are equipped with an accessory pigment called peridinin that imparts the golden-brown hue. Most species are motile and as the name suggests, motility is facilitated by two flagella. The flagella beat in two separate grooves, one groove encircling the cell near the middle, the other perpendicular to the medial one. Many species are also "armored," that is, they are made up of an arrangement of cellulose plates, the number and shape of the plates particular to genus and species. While those dinoflagellates that contain plastids are photosynthetic, some, like those without plastids, also prey on algae, ciliates, and other organisms.

Of the two types of dinoflagellates that I see today the more abundant is *Peridinium* (Figure 1.16). There are dozens of them. They are shaped like little helmets and they twirl when they swim, a consequence of the

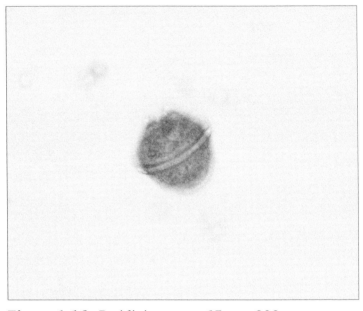

Figure 1.16. *Peridinium* sp. ~ 65 μm, 200x

positions of the flagella which act in unison. In fact, they move so fast that my eyes cannot keep up with them. I pick one individual and try to chase it by frantically manipulating the slide stage but it is a futile exercise; I simply cannot compete. As I continue to scan the sample I notice one specimen that has suddenly stopped moving. This is my one and only opportunity for a photo and I take it. (Yes, I could learn the techniques for collecting, killing, and making permanent mounts of specimens for photography, but all of my sensibilities argue against it.)

Most species of *Peridinium* occur in freshwater but there are marine species as well. One or two of these are pretty well-known as they can bloom into high densities that produce toxic "red tides."

Peridinium can be confused with another common dinoflagellate, *Gymnodinium*. However, close observation will reveal an absence of armor plates in *Gymnodinium*.

The second dinoflagellate in my sample is *Cystodinium* (Figure 1.17). Like *Closterium* it has an exquisite lunar shape but in this case the cell is enclosed in a sheath with extensions that appear as "horns." It also has a red eyespot. Unlike *Peridinium*, it is non-motile, a bonus to the weary observer. I wish I could know the species name of this particular individual but I find that in identifying protists it is generally very difficult to venture beyond the genus level. In the case of *Cystodinium*, for example, 33 species are presently listed but only 18 are accepted taxonomically. Protist species differentiation is frequently based on details of fine-structure only discernible by phase contrast or scanning electron

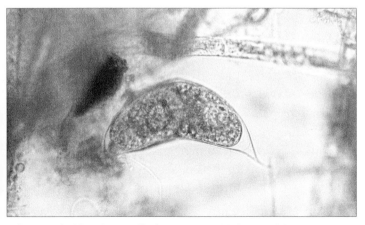

Figure 1.17. *Cystodinium* sp. ~ 115 μm, 200x

microscopy. Or, in some instances, characterization is dependent on a thorough knowledge of the organism's life-cycle. Molecular data are now helping to sort things out, but much also simply depends on consensus among specialists. As protist taxonomy seems fraught with difficulties, I retreat. I also comfortingly recall Walt Whitman's caution in connection with trying to be too precise about Nature: "*A certain free-margin, and even vagueness — perhaps ignorance, credulity — helps your enjoyment of these things. . .*"

In general, the *most* common components of my freshwater samples are diatoms. Not usually present in great variety as I have noticed for some other algal protists, they are nevertheless *always* present. While most species are unicellular, colonial forms are common too. Cell wall structure in this group is unique in that it comprises two valves made of silica known as a frustule. One valve fits neatly and securely over the other as a lid fits over the base of a little gift box. The cell's protoplasm is safely housed inside. Electron microscopy reveals an astonishing array of intricate ornamentation adorning frustules, the designs specific to each species.

Of the two basic shapes of diatoms, pennate and centric, what glides slowly into my field of view today (and most days) is a rather large, elongate, pennate type (Figure 1.18). Its color is deep gold, the result of two large plastids incorporating the golden-brown carotenoid pigment fucoxanthin which masks chlorophylls *a* and *c*. The interior of the cell is seemingly awash in oil. A product of photosynthesis, this lipid material is a form of reserve storage for the cell. While most diatoms are photosynthetic autotrophs, some species, when occurring

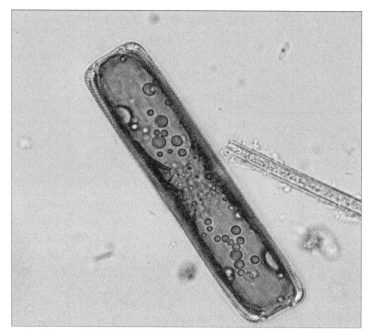

Figure 1.18. Pennate diatom *Pinnularia* sp. Note the oil droplets. Girdle view. ~ 130 µm, 200x

in environments with low levels of exposure to sunlight, become heterotrophs. In effect, they are able to live and grow in darkness by extracting organic carbon from their environment.

Because of their durable silica frustules, diatoms have been preserved in the fossil record since the Lower Cretaceous. Indeed, much of the world's crude oil supply derives from fossil diatom beds. Today they are found in both freshwater and marine habitats, the majority of species occurring in plankton. They are a major source of food for aquatic animals providing fatty acids, carbohydrates, and other nutrients. As such, there is no quantifying the extent of their world-wide ecological significance.

I feel obliged here to pay tribute to America's pre-eminent microscopist and first phycologist (algae specialist) Jacob Whitman Bailey (1811–1857). A graduate of West Point's class of 1832, Bailey was subsequently appointed professor of that institution's new department of chemistry, mineralogy, and geology. Spare time was devoted to botany and microscopy and, in 1839, these interests converged to induce what would become a life-long study of algae, principally diatoms. Bailey was a pioneer in this pursuit as very little attention had previously been paid to diatoms in America except as "test-objects" utilizing their complex frustule designs for determining the resolution powers of microscope objectives. In particular, they were used in an ongoing effort to demonstrate that the lenses crafted by the afore-mentioned Charles A. Spencer were as good as those of British and European makers such as Powell of London.

That he was plagued by chronic ill health and had lost his wife and daughter in the tragic explosion of the Hudson River steamboat *Henry Clay* in 1852, did not quell Bailey's passion for diatom research. It was, perhaps, a catharsis for the trauma in his personal life. In any case, he was able to muster the energy required to undertake difficult projects such as description of the diatoms collected on the Wilkes Expedition (1838–1842), the North Pacific Exploring Expedition (1853–1856), and, while suffering from a severe throat disorder, travel south during the winter and spring of 1849–1850 to search for specimens. Describing himself as "an invalid" at this time, he was yet able to extract both living and fossil diatoms from ditches, ponds, streams, rivers, rice fields, salt marshes, sulfur springs, sphagnum swamps, sea walls, and other sources while meandering through South Carolina, Georgia, and Florida over a seven-month period.

Bailey's pioneering work has largely been forgotten today but had he lived beyond the age of forty-six his prowess in diatom investigation may have rivaled that of Christian Gottfried Ehrenberg, Europe's premier protozoologist and Bailey's contemporary.

Monday, April 29*th*, 2:00 P.M.

Rugg Road Shrub Swamp. It has arrived – my 500 ml polyethylene beaker angled on the end of a three-foot handle — a *real* sampling device (Figure 1.19). Rugg Road is my most convenient water source so I have come here to try it out. I am now able to stand along the edge of the water and gently scoop up sample without actually stepping in, welcome relief to the obtrusion of wading in and trawling the sediment with my old Del Monte peach jar.

Back home as I begin to cruise microscopically through my initial slide of sample, the first resident of interest I encounter is not a protist. Included among Eubacteria, this is an olive-green, mucilaginous colony composed of chain-like filaments, a cyanobacterium called *Nostoc* (figures 1.20 and 1.21).

Nostoc is primarily a terrestrial cyanobacterium with blue-green pigments. It is an important nitrogen-fixer for liverworts and other plants, as well as a photobiont

Figure 1.19. Water sampling beaker.

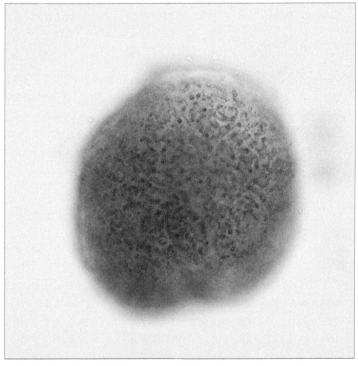

Figure 1.20. A typical, nearly spherical, mucilaginous colony of *Nostoc* sp. 100x

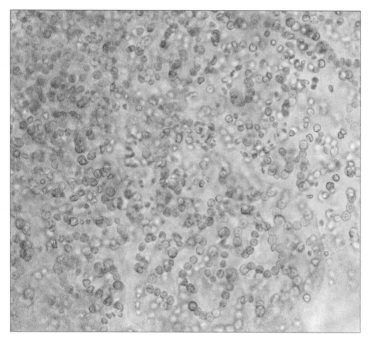

Figure 1.21. Close-up image of a *Nostoc* sp. colony showing coiled chains of spherical cells. 200x

(the photosynthetic food producer) in many species of lichens. I have encountered it before, macroscopically. In a damp corner of a local quarry it often occurs on the wet ground as round, flattened, olive-green colonies, roughly one inch in diameter, having the feel and consistency of Gummy Bears.

Nostoc is very similar to another common cyanobacterium, *Anabaena.* They both form similar chains of cells but those of *Nostoc* are embedded in mucilaginous colonies while those of *Anabaena* usually remain solitary.

I empty my concavity-slide into the "already examined" jar and refill it with sample. Almost immediately I

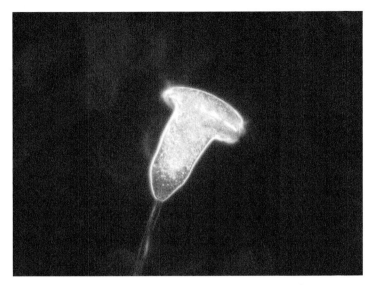

Figure 1.22. *Vorticella* sp. 100x, Dark Field

chance upon an old favorite, *Vorticella* (Figure 1.22). This ciliate protist is beautifully bell-shaped and clear. It is supported by means of a long stalk attached, at its base, to plant debris. As I gape, the stalk suddenly contracts with split-second speed and the oral cilia surrounding the rim of the "bell" retract. This impressive behavior is the organism's response to disturbance, in this case nothing that I, a bumbling human, have the subtle wit to perceive. Slowly the stalk unfurls, the bell expands, and cilia are thrust out once again. The cilia resume their beat creating visible currents that relentlessly draw potential food particles inward.

Ciliates like *Vorticella* are so named because of the cilia or hair-like projections borne on the outer cortex of their cells. They are among the most well-known protists and comprise the greatest number of protist species in some freshwater environments. Cilia perform two basic functions — locomotion and food gathering. Food for most ciliate species consists of bacteria and other protists.

Roughly 10,000 freshwater and marine ciliate species have been described and it is believed that many more exist, yet to be reported. Because most ciliates do not have hard parts that allow for fossilization, the fossil record for this phylum is poor. However, researchers using the "molecular clock" approach based on genetic divergence data, have estimated that the origin of ciliates *may* date to a very ancient 2 billion years ago.

Vorticlla was the first ciliate with whom I made acquaintance after acquiring my microscope and I was *stunned* by it. How could I have not heard of this marvelous organism before? Why had it never been included with the herons, frogs, and water plants on "Save Our Wetlands" T-shirts and posters? This indifference to micro-communities is egregiously uncircumspect. We, with our global economy, manic electronic connectedness, and yearning to find life elsewhere in the solar system, are, sadly, still strangers to two-thirds of life on our own planet.

Friday, May 3ʳᵈ, 11:45 A.M.

The Swamp. Historically, the word "swamp" invokes all sorts of malignant descriptions and images. But today, as I stand beside the multi-acre swamp near my home, the only word that comes to mind is "beautiful" (Figure 1.23). Sun slants across the water's unbroken surface of duckweed burnishing it to an almost glaring sheen. Tussock sedges, which only a few weeks ago embodied the brown wreckage of winter, have been transformed into thriving little oases of vibrant green.

Figure 1.23. The Swamp.

The scores of dead trees, some pitched at odd angles, today do not detract; the sun has annulled any feeling of desolation that they might otherwise incur. This is truly "no man's land" but it is home to nearly every other kind of life. I am held by its alchemy but also conscious of my intrusion into its precious stillness. I tiptoe forth and back for sample and hurry quietly away.

It is a mildly warm day, 65°. My swamp sample is an even warmer 68° but it was collected from the shallow water close to shore which heats up quickly. pH, the measure of acidity or alkalinity, reads a neutral and healthy 7. Most freshwater organisms prefer a pH above

30

6. When the value falls below 6 and acidity increases, whether from the natural effects of local geology and hydrology or from human effected agents, changes in freshwater communities begin to take place. At levels of 5 through 6, algal species diversity as well as biomass decrease. Further decrease in pH causes a corresponding decline in faunal diversity. Fortunately, all of the waters in my neighborhood maintain levels of 6 through 8 which allow for the great diversity of microlife that I routinely see.

Lots of life in the sample today but many of the organisms are simply too tiny for my objectives to resolve. They form a pulsing background against which the larger organisms dart, jump, glide, and twirl. It is impossible to imagine the exponent to which the total biomass of Earth's waters would be raised, based on just one field of this sample. It would be like contemplating the enormity of the universe, an exercise that wholly confounds the mind.

As I scan my slide the rather predictable size range of organisms I see is suddenly knocked out of kilter by an alarmingly large creature that jumps into view. It is not an ostracod this time, but an early larval stage of a copepod (Figure 1.24). Like ostracods, copepods are microcrustaceans (Figure 1.25). They may be free-living or parasitic and are usually the most common microcrustaceans found in quiet and open-water microcommunities. Native species range in length from 0.3 to 3.2 mm. While most appear to be rather dull gray or brown in color, some are ablaze in purple, orange, or red. In a few species, reserve food in the form of oil globules imparts the bright color.

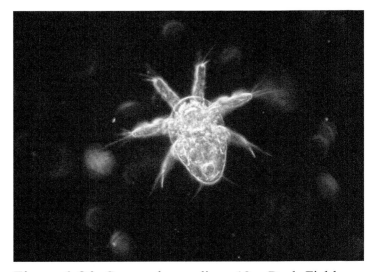

Figure 1.24. Copepod nauplius. 40x, Dark Field

The copepod body is elongate, segmented, and divided into three parts — head, thorax, and abdomen. The head and a portion of the thorax are covered with a carapace. Another conspicuous feature of the head is a pair of long, segmented antennae. These are sensory appendages but are also used for swimming. The length of the antennae vary, a fact that is useful for distinguishing among the three major groups of free-living individuals.

Food preferences for these little animals include protists, small metazoans, plant debris, insect larvae, and other microcrustaceans. Not surprisingly, copepods themselves comprise a vital link in the aquatic food chain, included among the prey of certain fish.

After fertilization, a copepod egg hatches into a lively larval stage called a nauplius. Additional nauplius stages follow, each stage increasing the length of the body.

Figure 1.25. Copepod, lateral view. 40x

Succeeding these are a number of so-called copepodid stages which finally result in the mature adult.

What has landed in my field of view today is an early-stage nauplius having three pairs of appendages and a prominent red eye. "Nauplius" was originally a genus name given to these larvae before it was understood that they are not mature animals.

The metamorphosis process which produces nauplii is not limited to copepods. Most other crustaceans develop from a similar series of larval and immature stages, each stage followed by a molt in which the carapace is shed to allow for new growth.

I resume my travel through the sample and am delighted to see the "tumbleweed" of the protist world, *Synura*, rolling into view (Figure 1.26). *Synura* is a lovely colonial flagellate alga, the rather large, golden-brown colony roughly spherical in shape and formed by many individual cells held tightly together at the center by their posterior ends. Each club-shaped cell bears two flagella of unequal length on its anterior end, and the cell surface

Figure 1.26. *Synura* sp. colony. 100x, Dark Field

is covered with overlapping silica scales. The scales are ornamented and arranged around the cell in a very organized pattern. This scale morphology is the primary character of species identification.

Predominantly a freshwater genus, *Synura* may be found in habitats as diverse as the polar regions and the tropics. Some species, particularly *Synura petersenii*, can form blooms, even under ice, that confer unpleasant odors to their surrounding water.

Synura overwinters or avoids desiccation by forming siliceous ornamented cysts. Due to the presence of durable silica, cysts, along with cell scales, are present in the fossil record, the oldest found in Cretaceous sediments roughly 100 million years old.

As I drive my sample back to The Swamp I pass the shrub swamp on Rugg Road. A shiny dark surface on a tussock sedge just catches my eye. What is it about turtles that draws us to them? Perhaps it is because, like *Synura*, they too have traversed the ages, but as *sentient* witnesses to Earth's alternating cycles of stasis and change. However, they hold the essence of this experience unyieldingly close, as on Delphic scrolls stored deep within the domed vaults of their shells.

Sunday, May 5th, 2:45 P.M.

Ten-Mile Creek. This pretty stream has its source in a local lake and then traverses much of the township from north to south before joining Catskill Creek in the adjacent county (Figure 1.27).

My approach to the water is made easy by a public fishing path, and downstream, in the distance, I do indeed see a figure in waders fishing for trout. As I head upstream I find that nearly all the rocks are coated with a brown, slimy-looking substance (Figure 1.28) that makes walking over them very tricky. I retrieve a couple of small rocks and scrape the material into a collecting jar along with a bit of water before heading home.

Under the microscope the brown substance turns, almost literally, into gold. There are thousands and thousands of golden pennate diatoms. Of particular beauty is *Meridion*, a colonial form (Figure 1.29). *Meridion* colonies are circular in shape with a central opening, rather like a wheel without an axle. The shape arises from the wedge-shaped cells growing against each other, face to face, held together by mucilage. Complete colonies are rather rare; what one usually sees are semicircular fragments.

I am reminded of Jacob Whitman Bailey's similar experience with *Meridion* as recorded in an 1843 report:

Figure 1.27. Ten-Mile Creek.

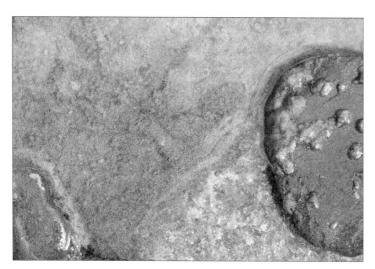

Figure 1.28. The brown, mucilaginous material on the rocks and in the water at Ten-Mile Creek.

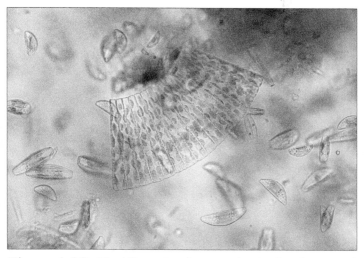

Figure 1.29. *Meridion circulare* against a background of other, assorted diatoms. 200x

"It occurs in immense quantities in the mountain brooks around West Point, the bottoms of which are literally covered in the first warm days of spring, with a ferruginous colored mucous matter, about one quarter of an inch thick, which, on examination by the microscope, proves to be filled with millions and millions of these exquisitely siliceous bodies."

Meridion, like many other organisms, has a distinct seasonality. While prolific now, its conspicuous presence on the rocks in the creek will soon be gone.

Wednesday, May 8ᵗʰ, 10:30 A.M.

Nordlund Pond. This is a small, nearby pond, very shallow, the water always dark from a high concentration of submerged plants and other organic material. A pair of Canada geese has made a nest just out of sight where a little stream enters the pond. I see one of the pair nearly every day as I pass.

In recent weeks the pond's surface has become host to many large, bright green, algal mats (Figure 1.30). The mats are Zygnematalean in composition, a complicated weave of *Spirogyra*, *Mougeotia*, and *Zygnema*. *Mougeotia* and *Zygnema* filaments are more narrow than those of *Spirogyra*, and the cells of *Zygnema* feature a pair of distinctive, star-shaped chloroplasts (Figure 1.31). *Mougeotia* is in the process of reproduction as evident by the presence of some conjugating filaments (Figure 1.32). Its flat chloroplast (Figure 1.33) can be rotated to accommodate changes in light so it appears differently in the cell depending on the plane of view (Figure 1.34).

Figure 1.30. Algal mats on Nordlund Pond.

In some cases pyrenoids, proteinaceous structures associated with the chloroplast, can be seen in a row within the chloroplast. This can be a useful diagnostic character.

Also present today, in more abundance than usual, are ribbons of the colonial diatom *Fragilaria* (Figure 1.35). Frustules are wider at their centers and held together there by tiny interlocking spines along the frustule margins. This species is probably *Fragilaria crotonensis*. It was first described in 1869 from sediment in New York City drinking water sourced from the Croton River via an aqueduct and reservoir.

Nordlund Pond is situated along the edge of a forest so pine pollen or "Pollen Pini" too is prevalent in the sample (Figure 1.36). Pine pollen was first described from American waters by C. G. Ehrenberg in his 1841 monograph,

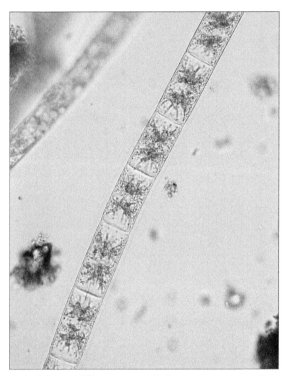

Figure 1.31. A filament of *Zygnema* sp. Note the stellate chloroplasts. 200x

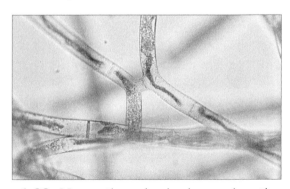

Figure 1.32. *Mougeotia* sp. beginning conjugation. 200x

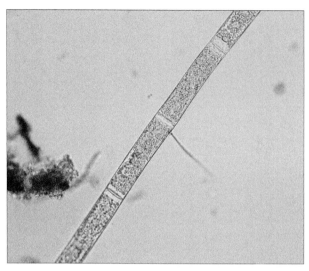

Figure 1.33. *Mougeotia* sp. cells with flat chloroplasts. 200x

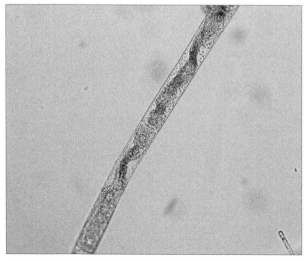

Figure 1.34. *Mougeotia* sp. with chloroplasts in two cells partially turned to take advantage of light. 200x

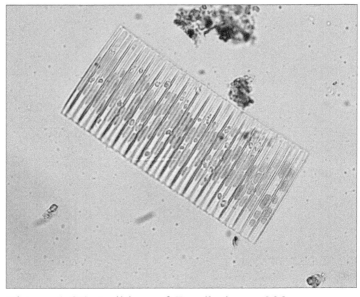

Figure 1.35. A ribbon of *Fragilaria* sp. 200x

On the Extent and Influences of Microscopic Life in North and South America. It was from perusing the lovely plates in this old volume that I was first able to identify pine pollen in my samples.

Saturday, May 18ᵗʰ, 1:30 P.M.

Even in the world of protists there are well-known celebrities. These are organisms like *Paramecium, Amoeba,* and *Volvox* whose names and "faces" habitually appear in textbooks and who play leading roles in high school biology demonstrations designed to excite students out of their enervating boredom. But *nothing* can prepare one for seeing *Volvox* out of character, in the wild.

Spring

Rugg Road Shrub Swamp. The day is overcast and calm. No turtles, but lots and lots of the green vegetative stems of *Equisetum* near the water's edge — so fine to see. These silica-rich plants have been a steadfast component of Earth's flora for millions of years.

Water temperature and ambient temperature are today, oddly, coequal. Polliwogs swim about in collected sample. I return it and scoop again. More polliwogs. I take the sample but will not keep it long.

In my first slide I see *Spirogyra, Synura, Closterium,* some other old acquaintances, and soon become lulled into the complacency of the familiar. But I should know better. As if someone had deliberately wound it tight

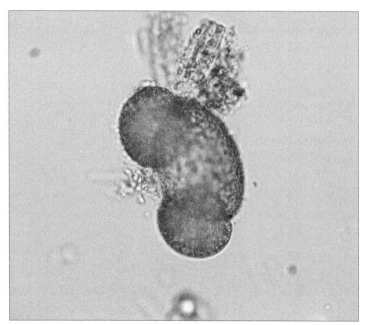

Figure 1.36. "Pollen Pini" or pine pollen. 200x

and then let it go, a colony of *Volvox* suddenly spins into view (Figure 1.37). It is shockingly large, shockingly green, and, shockingly beautiful. I am so surprised that for a moment I doubt what I see. Is *Volvox* not solely derived from biological supply house cultures? How did it get here? These thoughts are, of course, ridiculous but in all of my many forays into local freshwater habitats this is the first time I have ever encountered it. So ingrained is my association of this organism with the ideal of textbooks that it never occurred to me to expect it. As my heart begins to resume a normal beat I chase after the colony for photos; I need them to prove to myself that I have indeed actually seen it.

Everything about *Volvox* is extraordinary and it is no wonder that its first observer, van Leeuwenhoek, was so enthralled that he watched it for days on end. *Volvox* is a globe-shaped, hollow, coenobial colony that spins and rolls. Seen against the black background of dark field illumination it looks very like a tiny green planet. Motion is generated by coordinated beating of flagella, two each, on thousands of somatic cells. All these cells are embedded in a matrix layer surrounding the outside of the colony. As *Volvox* is a green alga, the somatic cells are photosynthetic. In addition, asexual colonies such as this one contain larger non-flagellated reproductive cells or gonidia (Figure 1.38). These cells undergo several mitotic divisions, eventually producing daughter colonies. Daughter colonies remain inside the parent colony with their flagella facing inward. However, at maturity they invert so flagella are directed outward. The mature colonies then release an enzyme that dissolves the parent colony's

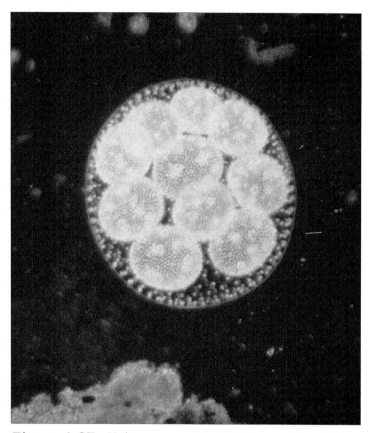

Figure 1.37. *Volvox* sp. parent colony with mature daughter colonies. Note the presence of gonidia in the daughter colonies. 100x, Dark Field

surrounding matrix and they are released to swim away (Figure 1.39).

Not only is *Volvox* an exquisite organism for the observer, it is also, by means of the ease with which it can be cultured, and the fact that it exhibits differentiation between somatic and reproductive cells, the

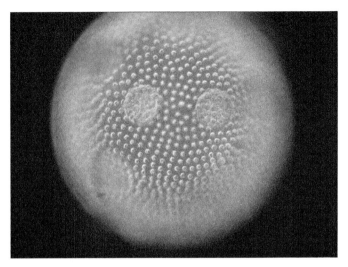

Figure 1.38. *Volvox* sp. colony showing somatic cells and gonidia. 200x Dark Field

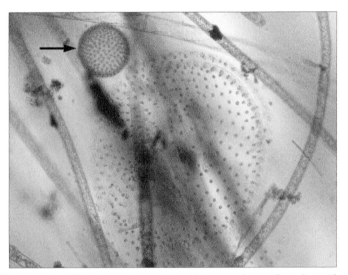

Figure 1.39. A mature daughter colony leaving the parent colony. 100x

perfect laboratory model. Indeed, *Volvox* has been used by molecular biologists to try to understand a very important step in the history of life — the evolution of multicellular organisms from unicellular ancestors.

Summer

Rugg Road Shrub Swamp. 73°. A perfect first day of summer. I collect some sample and meander homeward.

In the first slide I see the shelled or testate amoeba *Difflugia corona* (figures 2.1 and 2.2). This species was the first testate amoeba with whom I made acquaintance several years ago and at that time I was at a loss for means to identify it. Coincidentally, a local institution engaged in what has become the all too common practice of discarding old books, allowed some friends and me to rummage around in the bins of forsaken material before they were finally dumped. In a stack of U.S. Geological Survey reports from the 1800s I found Joseph Leidy's 1879 monograph, *Freshwater Rhizopods of North America*. I could not believe my good luck and immediately took it away, a glance thrown over the shoulder from time to time in case someone might try to trip me up. I need not have worried. Outside of Philadelphia and a few scientific circles Joseph Leidy's name is now virtually unknown. And while valuable to me, the volume had apparently been of little importance to the institution as evidenced by its nearly pristine, unused condition.

Joseph Leidy (1823–1891) was one of the most respected American scientists of the nineteenth-century. He is particularly noted for his work in vertebrate

49

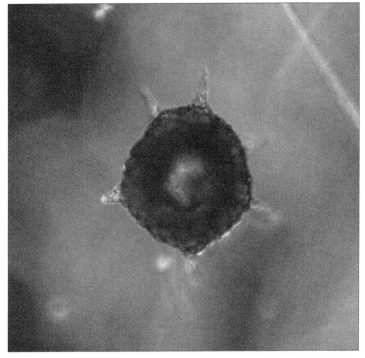

Figure 2.1. *Difflugia corona*. Note the six hollow spines. ~ 90 μm, 200x, Dark Field

paleontology and as a microscopist who made major contributions to parasitology and protozoology.

During the summers of 1872 and 1873 Leidy traveled west to Fort Bridger, the Uinta and Rocky Mountains, and the Salt Lake Basin to collect fossils and to concentrate on his rhizopod (amoeba) studies. He also collected amoeba specimens from ponds, springs, ditches, and marshes around his home area of Philadelphia as well as in New Jersey, Maine, Florida, and other localities.

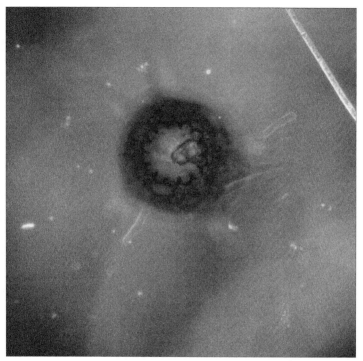

Figure 2.2. *Difflugia corona* showing 14 "teeth" around the aperture of the test. Pseudopodia are extended. ~ 90 μm, Dark Field

Leidy's monograph, quarto in size, comprises 324 pages of text and 48 color plates. All of the plates were delineated by the author and reflect an amazing understanding of the often difficult organisms he was observing. For precision of detail they are the "stacked" images of his day. He also includes descriptions of how he collected and maintained his amoeba specimens, sometimes keeping them alive in his study for an entire year. The monograph is still available in reprinted form, a valuable resource for nearly 140 years.

In the process of examining the monograph's plates I serendipitously found the identification of my amoeba specimen. Plate XVII clearly showed that it was *Difflugia corona*. I hadn't realized that testate amoebae were so common as textbook images have always favored "naked" ones. In fact, testate amoebae seem to be the more common local freshwater form of the organism. In the case of *Difflugia corona* the test is made of accreted quartz-sand particles acquired from its environment (as opposed to some other testate amoebae which secrete their tests). The amoeba resides within bottom sediment or on water plants in quiet-water environments, its food consisting primarily of algal protists. Leidy declares, *"Difflugia corona is the most remarkable and beautiful specimen of the genus."* It has come to be one of my favorite species as well.

Another amoeboid form in today's sample is one of the lovely "sun animalcules" or heliozoans — *Acanthocystis* (Figure 2.3). Its distinctive green color derives from the symbiotic green alga *Chlorella* residing within it. Heliozoa are characterized by a round naked body with radiating arms or axopodia. They are "passive predators," that is, they wait for prey such as ciliates, algae, flagellates, and small metazoans to approach and become entrapped in their axopodia. Small adhesive structures (extrusomes) on the axopodia hold prey for engulfment by the cell.

My sample is also visited today by two types of rotifers. These are metazoans, not protists. The name "rotifer" derives from early observers who thought the ciliated trochal disks on the head or corona of these little animals resembled rotating wheels. The vast majority of

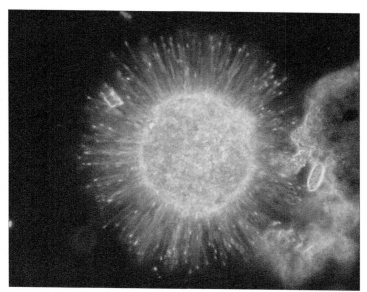

Figure 2.3. *Acanthocystis* sp. Body ~ 150 µm wide. 200x, Dark Field

rotifers reside in freshwater. They may be free-living, sessile, epizoic (living on the surfaces of other animals), or parasitic. Most are 100 µm to 700 µm long with the body usually divided into four regions — head, neck, body, and foot. These organisms seem to have some of the most bizarre body morphologies in the animal kingdom.

Of the two rotifers I see today, *Rotaria* and *Collotheca*, *Rotaria* is a bdelloid rotifer meaning that it creeps or crawls like a leech as it browses among algae or plant debris (figures 2.4 and 2.5). In contrast, *Collotheca* is sessile; it attaches to a substrate and pretty much stays put (Figure 2.6). Both rotifers' bodies are protected by secreted proteinaceous tubes that have the ability to telescope or contract when the animals are

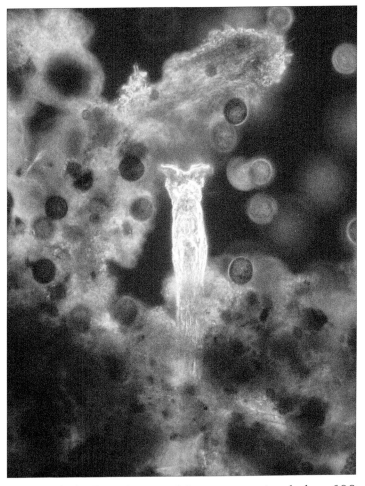

Figure 2.4. *Rotaria* sp. with corona extended. ~ 600 μm, 100x, Dark Field

disturbed. This action is facilitated by annuli of differing diameters encircling the body. They are also equipped with one to four "toes" on their feet, the toes possessing adhesive glands that allow the animals to attach

Summer

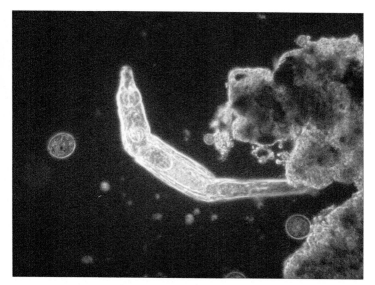

Figure 2.5. *Rotaria* sp. with corona retracted. ~ 600 µm, 100x, Dark Field

to substrates. The coronal cilia assist in locomotion and sweeping food particles into their mouths, their food preferences including algae, protists, organic matter, and smaller rotifers. Calcified structures called trophi are the only real hard part of these animals and are found as fossils. However, the protective tubes of *Habrotrocha angusticollis* have also been found preserved as fossils in *Sphagnum* peat samples of Holocene age.

Both *Rotaria* and *Collotheca* reproduce parthenogenetically. This mode of reproduction is particularly common in bdelloids, allowing them to thrive in the often unstable environments in which they are found. It has been suggested that ancient bdelloids may have reproduced sexually but abandoned this means over time in favor of obligate parthenogenesis as a survival strategy.

55

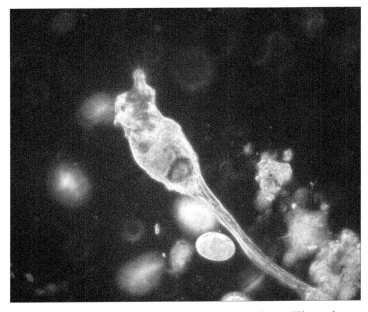

Figure 2.6. *Collotheca* sp., lateral view. The characteristic long setae on the corona are barely visible. Note the egg attached to the body by a filament. ~ 700 µm, 100x, Dark Field

Bdelloid rotifers also have the amazing ability to survive drought in a desiccated state of dormancy called anhydrobiosis. When rehydrated, they resume normal function. Even more astonishing is the fact that they do not age during their periods of dormancy. Dormancy may be months or a year in duration, but physiologically this time does not count in their overall age. The desiccated condition of these little animals also assists in their dispersal as they may be blown about for miles like particles of dust.

Wednesday, June 26*th*, 1:45 P.M.

Nordlund Pond. It is a warm day at 83°. Large bushes of *Rosa multiflora* thrive on the opposite shore, the air sweet with the scent of their white flowers.

My first slide is crowded with entangled filaments of algae. I scan slowly along the filaments noticing some very small bdelloid rotifers whose identification I do not know. I continue in this way for some time, the ravel of material giving the feeling of picking my way through a very dense thicket. However, the payoff for this arduous journey through the filaments is arrival at a "clearing" and an encounter with a stalked specimen of *Gomphonema* (Figure 2.7). *Gomphonema* is a genus of pennate diatoms having over 400 taxonomically valid

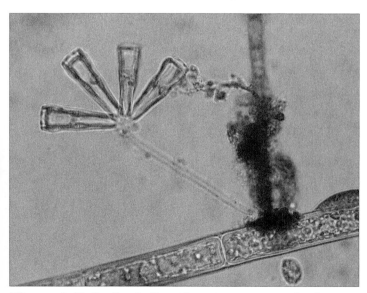

Figure 2.7. Stalked *Gomphonema* sp. attached to an algal filament. 200x

Figure 2.8. Stalked *Cymbella* sp. 200x

species. What I have chanced upon today is four wedge-shaped frustules (seen in girdle view, the region where the two frustules overlap) atop a long, gelatinous stalk. The stalk is attached to an algal filament. This species of *Gomphonema* is rather an outrageous thing in its appearance, but it is not unique. *Cymbella*, another diatom common in my samples (Figure 2.8), also appears from time to time in a similar stalked form, sometimes in company with *Gomphonema*. Stalks allow these diatoms to extend themselves above the rest of the diatom population giving them advantage in the competition for light and nutrients. Some diatoms, like *Cymbella*, also form gelatinous tubes attached to rocks in which they align themselves (Figure 2.9). This habit may be a defense against predators or it may simply afford a way to outcompete other epilithic diatoms.

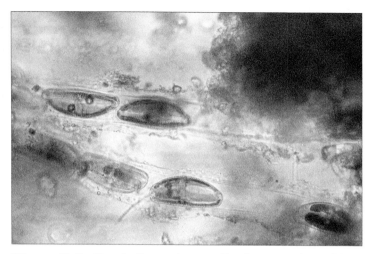

Figure 2.9. *Cymbella* sp. in mucilaginous tubes. 200x

As I continue to scan the slide I find the rotifer *Testudinella* seemingly trapped among some thick plant debris although it does not appear to be in distress (Figure 2.10). Its coronal cilia are beating at full tilt with material continuously drawn into the mastax. This organ is specific to rotifers and consists of muscles that control the jaw-like trophi. It is here that maceration of food occurs before it passes on into the rest of the digestive tract.

My next slide is much less crowded and I see a fine example of *Eudorina* (Figure 2.11). As *Eudorina* resides in the same taxonomic family (Volvocaceae) as *Volvox*, it too is a coenobial colony or hollow sphere incorporating a specific number of cells, in this case 16, 32, 64, or 128. Cells reside a little distance from each other, peripherally, inside an envelope of mucilage. Cell shape is ovoid

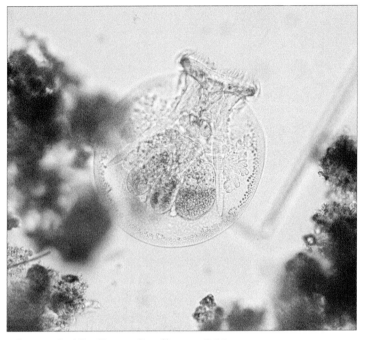

Figure 2.10. *Testudinella* sp. 200x

or spherical and each one disports two flagella of equal length. The flagella are parallel as they pass through the mucilaginous envelope, but then diverge, their coordinated beating allowing the colony to move about. The colony I observe today contains 32 cells and its protective envelope shows up nicely under bright field illumination.

Also present in this bit of sample is the beautifully deep green desmid *Pleurotaenium* (Figure 2.12). It is a large cell with a length exceeding 800 μm.

I prepare another slide and in no time at all a cladoceran moves herky-jerky across my field of view. It looks a little odd and I chase after it to find that it is

Summer

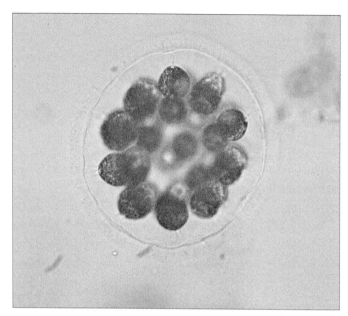

Figure 2.11. *Eudorina* sp. Note the mucilaginous envelope around the cells. 200x

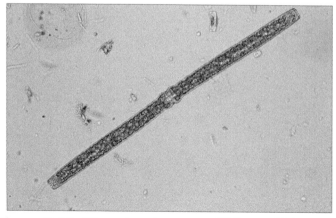

Figure 2.12. *Pleurotaenium* sp. ~ 825 μm, 100x

61

playing host to what appear to be many tiny peritrich epibionts (figures 2.13 and 2.14).

Cladocerans are a group of microcrustaceans commonly referred to as "water fleas."

Most are 0.2 to 3.0 mm long and their bodies are covered with a carapace that rather appears to be bivalved when viewed ventrally but, in fact, is one piece with a cleft. The carapace can be variable in shape and often has surface ornamentation. However, the most noticeable body feature of these animals is a large, dark, compound eye.

Cladoceran food consists of organic detritus and bacteria, hence they are helpful in keeping water clean. They, in their turn, are an important food for many fish.

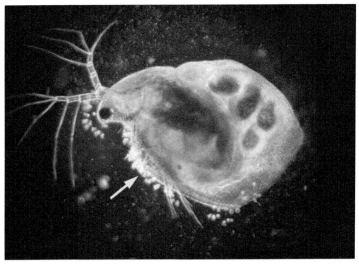

Figure 2.13. *Simocephalus* sp. cladoceran. Note the epibionts on the carapace and embryos in the brood pouch. 40x, Dark Field

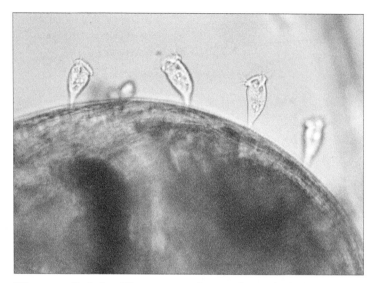

Figure 2.14. Close-up view of peritrich ciliate epibionts on a cladoceran. 200x

I have encountered cladocerans many times before in my samples but never with epibiont "passengers." No sample is ever routine — there is always something new.

Saturday, July 6th, 12:45 P.M.

Ostracod Ditch. A beautiful summer day, the fields filled with a profusion of Cow Vetch, Yellow Bedstraw, Bird's-foot Trefoil, and Black-eyed Susans.

My acquaintance with naked amoebae has always been minimal so at the outset of today's collecting foray I decide to make a deliberate search for them using Joseph Leidy's method:

"For ponds, ditches, or other waters I use a small tin ladle, or dipper. . . The dipper is used by slowly skimming the edge along the bottom of the water so as to take up only the most superficial portion of the ooze, which is then gently raised from the water and transferred to a glass jar."

My "dipper" is, of course, my plastic sampling beaker and I place my recovered sample of surface ooze (or biofilm in today's parlance) in a 3-inch by 3-inch jar. I allow the mud to settle and then pour off most of the water. The results are thrilling! In addition to five green *Difflugia* (figures 2.15 and 2.16), seven clear *Difflugia* (Figure 2.17), and a couple of other testate amoebae of

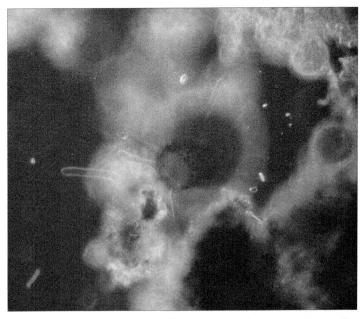

Figure 2.15. Green *Difflugia* sp. with pseudopodia extended. The green color derives from an algal endosymbiont. 200x, Dark Field

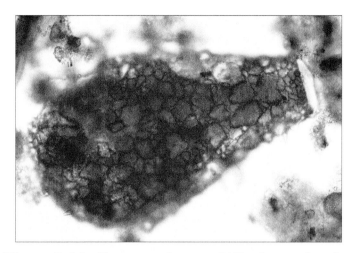

Figure 2.16. Close-up of green *Difflugia* sp. showing quartz crystals comprising the test. ~ 160 μm, 100x

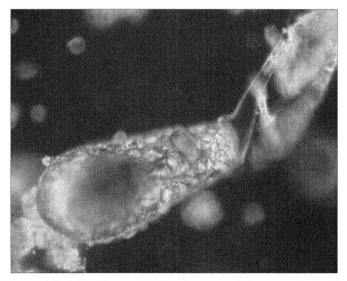

Figure 2.17. Clear *Difflugia* sp. with pseudopodia extended. 200x, Dark Field

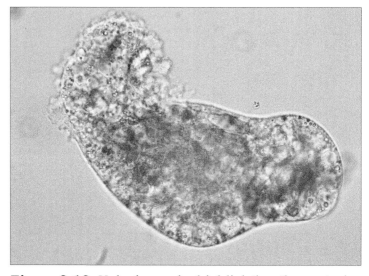

Figure 2.18. Naked amoeba highlighting the posterior uroid. ~ 225 μm, 200x

uncertain identity, I find eight naked specimens. They all appear to be the same species, that is, they are monopodial, have uroids (the bulbous posterior areas of moving amoebae), cytoplasmic crystals, rounded anterior edges that exhibit clear, fluid cytoplasm (hyaloplasm), and are roughly 225 μm in length. (Figure 2.18).

Identification of naked amoebae is difficult and dependent upon several characteristics including number and shape of pseudopodia, type of uroid, presence or absence of cytoplasmic crystals, shape and number of nuclei, etc. It is also necessary for the amoeba to be in locomotion, not in a resting state, for accurate evaluation.

Naked amoebae, it turns out, are not all really "naked." Recent SEM photographs have revealed that some are covered with scales of different shapes and intricacy, particular to species. This discovery adds yet another level of intrigue to these marvelous cells, information that Joseph Leidy would have truly appreciated.

Friday, July 12th, 11:15 A.M.

Willow Brook. This is the charming old-map name of the stream that runs past my home. In the spring the water is clear and swift, Trout Lily is abundant in the woods along its course, and the eye-catching liverwort *Conocephalum* can be found in areas of shade. Today the water level is so low that only a few pools remain — the summer face of Willow Brook. I clamber down a weedy bank to one of the larger pools and notice that much of it is filled with olive-green clouds just beneath the surface (Figure 2.19). I scoop "Sample A" of this and another, "Sample B," from among the rocks where the water is clear.

As always under the microscope, all is revealed. The dusky clouds of Sample A become beautiful branched filaments of *Cladophora*, a green alga common in streams. And there is a bonus. Covering the filaments like golden-brown ornaments are frustules of the centric diatom, *Cocconeis* (Figure 2.20).

Sample B shows *Melosira*, a centric diatom whose cylinder-shaped cells are joined in chains by mucilage (Figure 2.21). Gliding energetically among them is an odd looking little animal with scales on its dorsal side and many rather long bristles (Figure 2.22). It is the gastrotrich *Chaetonotus*, a very common resident in all of my water sources. Gastrotrichs glide by means of cilia on their ventral sides. They browse, feeding on algae, bacteria, and organic detritus, and are themselves consumed by a host of predators.

Figure 2.19. "Clouds" of *Cladophora* sp. in Willow Brook.

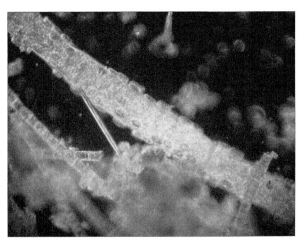

Figure 2.20. *Cladophora* sp. filament with attached *Cocconeis* sp. diatoms. 100x, Dark Field

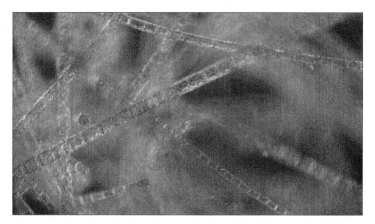

Figure 2.21. Chains of *Melosira* sp. diatoms. 100x, Dark Field

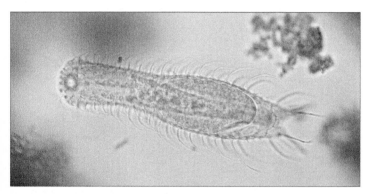

Figure 2.22. *Chaetonotus* sp. ~ 300 µm, 200x

Diatoms, as mentioned above, are extremely useful as indicators of stream and river health as the diversity and density of their populations are closely related to levels of pollution. This concept was first put into wide-scale use by Dr. Ruth Patrick (1907–2013) of the Academy of Natural Sciences in Philadelphia in the 1940s.

She was a pioneer in environmental research and spent sixty years monitoring river systems in the United States, advocating for clean water. The Academy of Natural Science's diatom collection, much of it contributed through Patrick's efforts, is the second largest in the world. In perusing the online catalogue I am delighted to discover specimens within it that were collected from Ten-Mile Creek, the local creek into which my little Willow Brook flows.

Thursday, July 18th, 2:30 P.M.

Rugg Road Shrub Swamp. Hot and humid 91°. The water level here is much lower now, the result of continuing high temperatures without rain. Mats of *Riccia fluitans* cover much of the surface and I find tiny "fingernail clams" (*Musculium securis*) on some of them. The temperature of my water sample is a very warm 86°.

I set up my first slide. What a crowd! There are organisms everywhere! Algae, rotifers, ciliates, gastrotrichs, testate amoebae, a newly-released colony of *Volvox*. . . The heat, concentration of organic material, and abundant prey have driven population diversity and density to a peak. I realize that of all my sample sources this small, unassuming, shrub swamp is literally the hotspot in terms of micro-diversity. I also realize that I have only begun to scratch the surface of becoming familiar with micro-communities. My samples derive from muddy shoreline areas or shallow streams. I do not have the means to plumb the deeper water benthos or sweep planktonic surfaces.

An interesting paper in the Summer 2013 issue of the *Quekett Journal of Microscopy* describes a 22-year

(1991–2012) biodiversity survey of a lake in Dorset, England, focusing exclusively on algae. Sampling of benthic and planktonic algae was performed on a monthly basis. Access to a boat allowed for sampling beyond the shoreline and in the center of the lake. More than 80 genera of planktonic algae and nearly as many from benthic samples were recorded. Chemical analyses were also conducted to monitor water quality over the survey period. The author, a retired non-specialist in algae, undertook the study, in part, as a form of recreation. However, as no systematic microscopic survey of the lake had ever been initiated, his dedicated work has become a valuable record. While molecular biology methods, which allow for recovery of all DNA in a given sample and subsequent genome sequencing to determine what microorganisms are present, may outdate this sort of hands-on biodiversity analysis, there is still a need for dedicated microscopists such as this one. For example, new species among protist groups, particularly ciliates and amoebae, still await discovery.

From among the "crowd" in today's sample I have the good fortune to find *Chlorohydra* (*Hydra viridis*) in my third slide (Figure 2.23). And what an impressive sight it is! As a consequence of the green alga *Chlorella* living as an endosymbiont in enzyme-resistant vacuoles among its cells, the entire hydra body is bright green. *Chlorohydra* benefits nutritionally from the presence of *Chlorella*, specifically from maltose, a by-product of photosynthesis, that seeps from the algal cells into those of the hydra. However, it is unclear how the algal cells benefit from the relationship except, perhaps, that they are safe from predators.

Figure 2.23. The tentacles of *Chlorohydra (Hydra viridis)*. 40x, Dark Field

Chlorohydra has a very simple body plan; it is basically a sac with tentacles surrounding an oral aperture. It is, however, a metazoan as it is composed of distinct cell layers having specialized functions. Food consists of prey such as microcrustaceans that come in contact with the tentacles and are stung by structures called nematocysts containing neurotoxin. The paralyzed prey item is then moved by the tentacles to the mouth which opens into the sac for digestion.

When the body is fully extended, *Chlorohydra* is roughly 3 mm long. In addition to *Chlorohydra*, brown hydras are also common and, like *Chlorohydra*, may be found on a variety of substrates including plant debris, submerged stones, and aquatic plants.

Contrary to most life on Earth, *Hydra* are seemingly immortal. Because of their ability to regenerate all body parts they never die of injury. They *will* die from exposure to heavy metals and detergents, and will succumb from intense starvation and extreme changes in water temperature. However, *Hydra* living in unpolluted water under stable conditions never show any signs of aging.

Certain species of *Chlorella* are the most habitual algal endosymbionts in protists and micro-metazoans. This ubiquity is likely due to the makeup of their cell walls which help drive formation of perialgal vacuoles in their hosts. In addition to *Hydra,* the heliozoan *Acanthocystis*, and some flagellates, *Chlorella* are most commonly observed in ciliates. I see a beautiful example of one — *Euplotes daidaleos* — today (Figure 2.24). *Euplotes* is one of the best-known ciliated protists as it has a habit of "walking" on bristles or cirri on its ventral side (Figure 2.25). It is omnivorous and is usually found foraging among submerged vegetation and plant debris. While the green form is most frequent in my samples, I have noted clear ones as well.

My last slide of sample of the day holds yet another interesting resident, the ciliate *Epistylis* (Figure 2.26). Like *Vorticella*, it is a peritrich ciliate bearing an oral ring of cilia, but the cell is more elongate in shape. It is a colonial form, the colonies joining to form dichotomous branches. *Epitsylis* is anchored to a colony of *Nostoc* and is a pretty contrast to today's profusion of relucent, *Chlorella* green.

Saturday, August 9ᵗʰ, 12:30 P.M.

The Swamp. A gorgeous summer day. I am on a mission — to find photosynthetic euglenoids. The

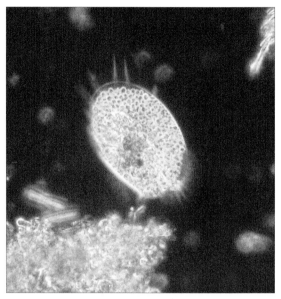

Figure 2.24. *Euplotes daidaleos*. Note the *Chlorella* endosymbiont. ~ 95 µm, 200x, Dark Field

Figure 2.25. *Euplotes daidaleos* "walking" among plant debris. 200x, Dark Field

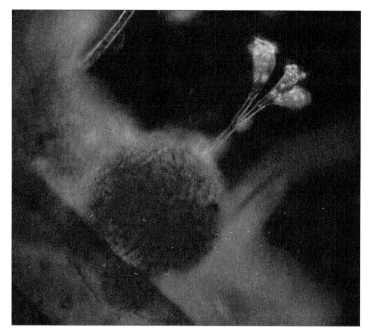

Figure 2.26. *Epistylis* sp. attached to a colony of *Nostoc*. 200x, Dark Field

Swamp is a reliable source for them, especially this time of year.

Euglenoids, or, more formally, Phylum Euglenozoa. How does one do them justice in words? As a group, most are distinguished by the surface of their cells which are called pellicles. These are composed primarily of proteinaceous "pellicle strips" that are parallel and arranged either helically or longitudinally and help define the way the organisms move.

The presence of paramylon bodies in photosynthetic euglenoids is also a very distinctive feature. Paramylon is a storage product in the form of a membrane-bound

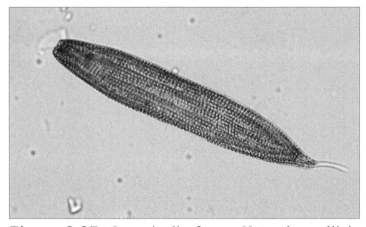

Figure 2.27. *Lepocinclis fusca.* Note the pellicle ornamentation. ~ 156 μm, 200x

crystal. Depending on the species, paramylon bodies may be present as discs of varying shape and size, rods, plates, or links. Paramylon bodies are usually visible with a light microscope and are a very useful differentiating character.

Euglenoid movement is generated by an anterior flagellum, and located near its base is an eyespot or photoreceptor. Eyespot size varies as does the intensity of its red color.

The joy of watching euglenoids is in the variety of their shapes and movements. They glide, they twist, they writhe, and there seems to be no end to their different shapes, a quality made even more interesting by the ability of many species to change them (metaboly).

As hoped, euglenoids are present in today's sample. Most conspicuous is a long, cylindrical, brownish-green individual with pellicle ornamentations. This is *Lepocinclis fusca*, the species name referring to its brownish color (Figure 2.27).

Next, *Phacus longicauda* glides into view (Figure 2.28). This organism's name describes its most salient feature — a long cauda or "tail." It also has a rather large paramylon disc, centrally located, and a dark red eyespot. Longitudinal striations are easily visible on the flat, leaf-shaped body of the cell.

Lest I forget the great diversity of their cell architecture, a very spiny euglenoid suddenly appears. This is probably *Trachelomonas armata* (Figure 2.29). Genus *Trachelomonas* is characterized by the presence of a mineralized sheath or lorica around the cells. An opening at the anterior end allows for the emergence of the flagellum and in some species this opening is surrounded by a clearly defined neck or collar. As in the case of *T. armata*, the lorica is often ornamented with spines of varying

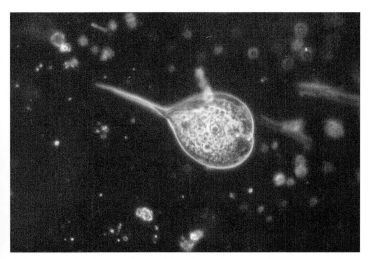

Figure 2.28. *Phacus longicauda*. Note the pellicle striations, paramylon bodies, and red eyespot. ~ 180 μm, 200x

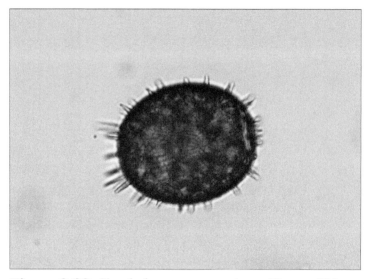

Figure 2.29. *Trachelomonas armata.* ~ 40 μm, 400x

shape and length. Ridged and punctated lorica surfaces
are common as well. *Trachelomonas* are not usually all-
green in color. More often they are distinguished by a
reddish hue which may derive from iron, one of the
chemical components of the lorica. Cell shapes range
from spherical to ovoid, but a few fusiform and ellip-
soid forms occur too.

I am able to identify most all of the euglenoids I see
to species level by means of Ciugulea and Triemer's
marvelous volume, *A Color Atlas of Photosynthetic Euglenoids*. The photographs and descriptive material in this
work are outstanding. One wishes that references such
as this were readily available for more groups of protists
and microscopic metazoans as they would go a long way
to exciting interest, and to resolving identification issues
that standard line drawings simply cannot do.

Summer

As a bonus in today's sample I see a very large rotifer encountered on rare occasions. It is squarish in shape and has spines on both ends. The body is encased in a clear lorica and the little animal moves quickly with a rolling motion, sometimes with a large egg attached in a manner similar to *Collotheca*. This is *Platyias patulus* (Figure 2.30). Its size is exceeded by another species, *Platyias quadricornis* (Figure 2.31), which I have also observed in samples from The Swamp. Both species are wonderful to see for their striking morphologies.

Paramecia abound today. They are such well-known ciliates and yet I rarely see them. I have observed more

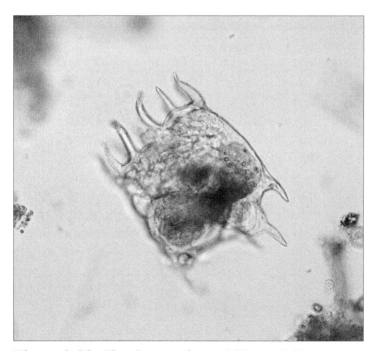

Figure 2.30. *Platyias patulus.* ~ 180 μm, 100x

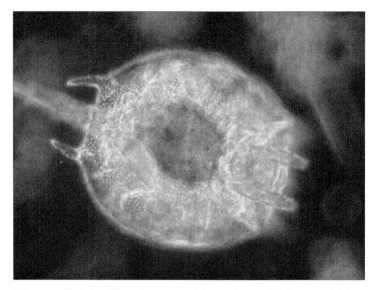

Figure 2.31. *Platyias quadricornis.* ~ 250 μm, 100x, Dark Field

today than perhaps ever before, including *Paramecium bursaria* (Figure 2.32). This is a beautiful ciliate, rather large at roughly 150 μm, and brightly green with a *Chlorella* endosymbiont. *Paramecium bursaria* is the only species of *Paramecium* that can maintain an endosymbiotic relationship with *Chlorella* for reasons that are, as yet, unknown. It glides gracefully among the plant debris in search of prey. However, it has the ability to accelerate to high speed seemingly without provocation and it does so; in an instant it is gone.

Monday, August 11th, 1:15 P.M.

Ostracod Ditch. Another lovely summer day but warmer with a water temperature of 84°. The ditch

Summer

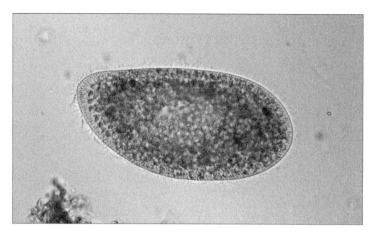

Figure 2.32. *Paramecium bursaria*. ~ 150 μm, 200x

has retained water all summer so far, a rather unusual condition.

First slide: Teeming with life — a *Trachelomonas* euglenoid swarm (Figure 2.33). There are thousands of them, their flagella clearly visible and whipping madly. They are probably *Trachelomonas hispida* var. *duplex* as the cells are elliptical in shape, have eyespots, and sport stubby little spines on both ends. However, they move too fast to determine paramylon size and shape.

With a population of prey this dense some predators must surely be lurking and in the very next slide I spy a naked amoeba engulfing one of the euglenoids.

Ostracods are again present here and in large numbers. A beautifully green-banded individual comes to a screeching halt precisely in the middle of my field of view (Figure 2.34). No flailing. Would that this kind of cooperation were not a one in a million occurrence!

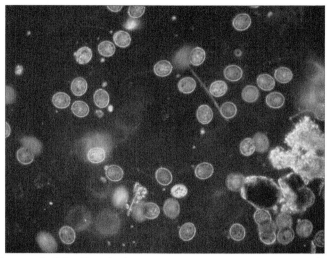

Figure 2.33. *Trachelomonas* swarm. 200x, Dark Field

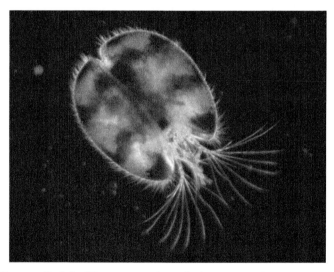

Figure 2.34. The green-banded ostracod. 40x, Dark Field

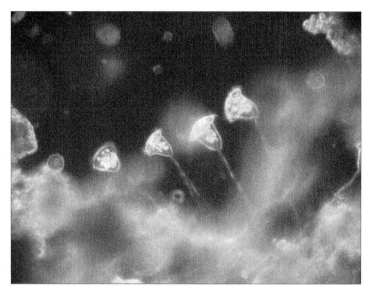

Figure 2.35. *Vorticella* sp. "bell choir." 100x, Dark Field

The *Trachelomonas* continue to move frenetically in every slide and I eventually become weary from watching. Just as I am about to remove my last slide from the microscope I notice some *Vorticella* (Figure 2.35). There are four, all in a row, with stalks extended. A very tiny bell choir.

Saturday, August 17th, 12:45 P.M.

Rugg Road Shrub Swamp. Last week's heavy rain has brought the water level back to nearly normal.

Three examples of an old algal favorite in today's sample — *Micrasterias* (figures 2.36, 2.37, and 2.38). This is a desmid genus with an abundance of species. Exhibiting an amazing diversity of cell morphologies,

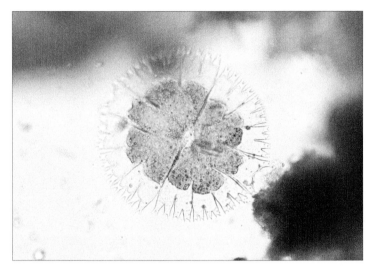

Figure 2.36. *Micrasterias* sp. ~ 180 µm, 200x

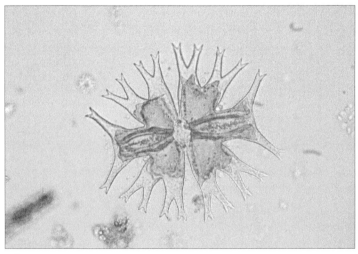

Figure 2.37. *Micrasterias* sp. ~ 125 µm, 200x

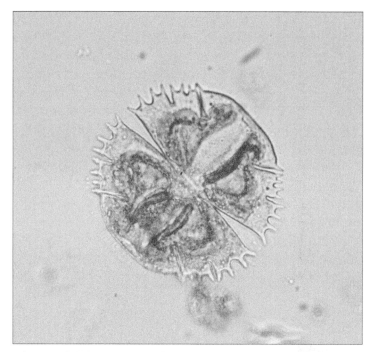

Figure 2.38. *Micrasterias* sp. ~ 120 µm, 200x

Micrasterias encompasses some of the most intricately beautiful of all protists.

Micrasterias brings to mind the rather long-standing academic debate over the distribution and biogeography of protists. Researchers in the "ubiquity" camp have maintained that very large population sizes and easy dispersal of protists must necessarily ensure global distribution on an "everything is everywhere" basis. However, the "endemism" camp has proposed that, while protists do have a wider range and distribution than macroorganisms, some protists are indeed endemic or restricted to particular areas. Once again, molecular analyses are providing important new data that, in this case, seem to quell the debate by showing that some protists are cosmopolitan while others are restricted. *Micrasterias* is an example of a genus that can demonstrate restricted distribution as some species of this genus have been shown to be confined to very specific parts of the world.

Another green protist in today's sample is the colonial alga *Pandorina* (Figure 2.39). The colony is ovoid in shape and contains 8 or 16 keystone-shaped cells within a mucilaginous matrix. Each cell supports two flagella of equal length.

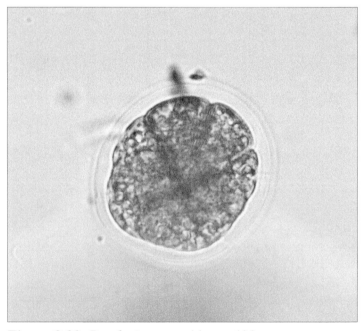

Figure 2.39. *Pandorina* sp. ~ 44 μm, 400x

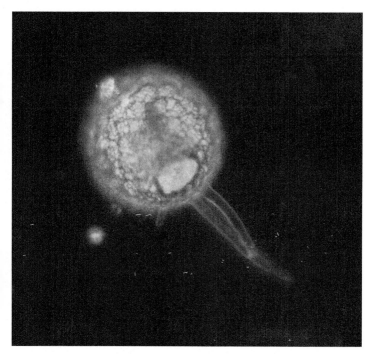

Figure 2.40. *Difflugia globulosa* with pseudopodia extended. ~ 95 µm, 200x, Dark Field

Amoebae too are present. A naked individual appears to have recently engulfed a meal, and a testate form, probably *Difflugia globulosa* (Figure 2.40), seems to be searching for food with pseudopodia extended.

Friday, August 23ʳᵈ, 9:00 A.M.

Lawson Lake. Last evening the local television news reported closure of the public swimming area at nearby Lawson Lake because of a toxic bloom. I am curious and decide to take a look. Even as I approach the lake

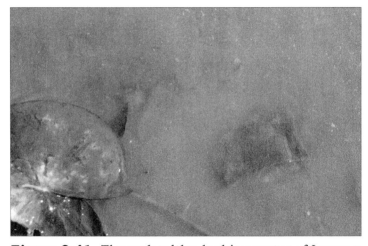

Figure 2.41. The unhealthy-looking water of Lawson Lake.

on the road I can see a definite green tinge to the water. Standing at the water's edge reveals a very murky bright green color (Figure 2.41) with quite an unpleasant odor. I sample from two separate areas and head home.

What appears under the microscope is a thick stew of cyanobacteria dominated by irregularly-shaped colonies of *Coelosphaerium* (figures 2.42 and 2.43). *Anabaena* filaments are also numerous and *Microcystis* too may be present. All three of these genera are often responsible for nuisance blooms in eutrophic waters and all three are known to produce toxins. Whether toxins have actually been detected in the lake water by environmental officials does not seem to be much of an issue in light of the rather startling green turbidity and rank smell. Both are alarming enough to keep animals and people away.

This bloom in Lawson Lake is a problem in miniature. What is now occurring in western Lake Erie is

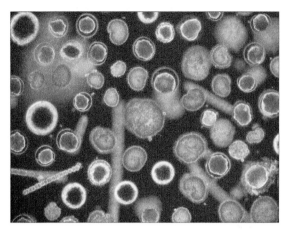

Figure 2.42. *Coelosphaerium*-dominated cyanobacteria from Lawson Lake. 200x, Dark Field

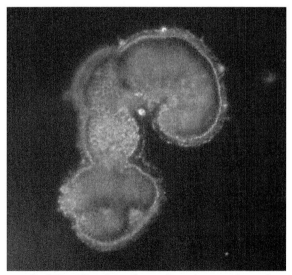

Figure 2.43. *Coelosphaerium* sp. showing the characteristic irregular shape of the colony, dense outer edge, and surrounding layer of mucilage. 200x, Dark Field

Lawson Lake on a grand scale. An intense bloom of *Microcystis* has been driven against the Ohio shoreline by winds and currents in the area of the water-intake for the city of Toledo. The levels of toxin — microcystin — produced by the bloom is three times higher than the allowable limit in drinking water. Unable to remove it, officials have had to close schools and businesses because of the lack of potable water. The situation is not new, but this year's bloom is particularly alarming, affecting more than 400,000 people. And the problem affects aquatic life as well. Decomposition of *Microcystis* by bacteria (which releases the intracellular toxin) can use so much dissolved oxygen that areas of the lake become "dead zones" where fish and other organisms cannot survive.

Climate change is increasing the temperature of the Great Lakes as well as the oceans. This condition, in combination with high concentrations of phosphorus from agricultural watersheds draining into Lake Erie, is primarily responsible for the production of large cyanobacteria blooms. A history of farmers applying fertilizers to frozen ground because of difficulties encountered with clay soils in the spring contributes much of the phosphorus. Not having the chance to be absorbed by soil, phosphorus-containing fertilizers are carried into watersheds by the spring thaw of snow and ice. Fortunately, programs are now being implemented in the Midwest to address fertilizer application issues and run-off management.

Water is the substratal medium of *life,* the numinous phenomenon of planet Earth. Here, there is *nothing* more important than water.

Sunday, August 31st, 2:00 P.M.

The Swamp. Rather a cool, late-summer day at 72°. Muggy, overcast, and showery. The measured temperature of the water is 71°.

I sample along the edge of the swamp suddenly realizing that I have forgotten my roll of pink flagging for assisting with the sample's return. I leave a Kleenex impaled on a shrub branch near my collecting point instead.

Perhaps it is the day's insipid weather that makes me feel a little dull for I am wholly unprepared for what I see in the first slide. There are *thousands* of photosynthetic euglenoids comprising at least ten species, all moving skimble-skamble in a green and reddish riotous mass (Figure 2.44). It is spectacular! I am all thumbs

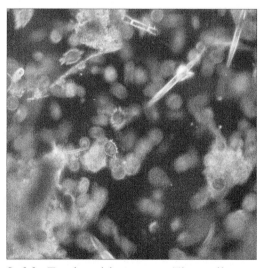

Figure 2.44. Euglenoid swarm. The cells are blurry in the photo because of their frenetic movement. 100x, Dark Field

trying to chase down individuals for photographs, their relentless movement allowing me to capture only a few. Frustrated after endless attempts, I just sit back and enjoy the show.

Trachelomonas armata, Lepocinclis fusca, and *Phacus longicauda*, previously noted, are all here. But there is a marvelous diversity of others as well. There is *Phacus gigas*, an ovoid cell with longitudinal pellicle striations, an angled cauda, and several small paramylon discs (Figure 2.45). *Phacus tortus*, similar to *P. helikoides* but smaller. Two species of *Lepocinclis* — *L. acus* and *L. helicoideus* (Figure 2.46) — both with elongated cells and rod-like paramylon bodies. However, *L. helicoideus* is larger and its body appears twisted as it swims. There is also a small, round *Trachelomonas* species, along with another that is somewhat elongate in shape and has a prominent flagellar collar. There may be other species as well, but the sample is so densely packed with moving cells that I can make no further determinations. If one were inclined to imagine euglenoids as much of a muchness, the reality of their wonderful diversity would be hammered home today.

After several hours of observation I drive the sample back to The Swamp but am in such a bleary daze that I just barely remember to remove the Kleenex from the bush as I leave.

Saturday, September 6th, 1:30 P.M.

Rugg Road Shrub Swamp. A gorgeous day — clear, sunny, and calm. The Catskill Mountains to the south are a lovely deep-denim blue.

I am still reeling from last Sunday's euglenoid extravaganza so my first slide today seems, at first, a little

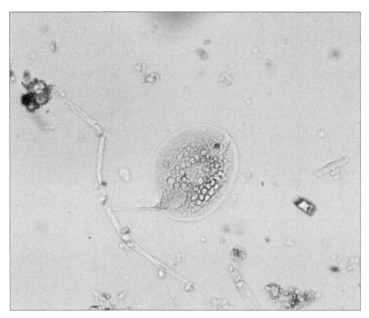

Figure 2.45. *Phacus gigas.* Note the numerous paramylon bodies and red eyespot. ~ 110 µm, 200x

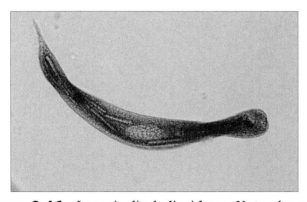

Figure 2.46. *Lepocinclis helicoideus.* Note the rod-shaped paramylon bodies and red eyespot. ~ 200 µm, 200x

sparse. Once I am settled in, though, it is actually pleasantly normal. *Chlorohydra* appears in all of its emerald green splendor, along with several other familiar forms. *Difflugia* amoebae are abundant as are *Arcella*, with brown-colored tests. The brown color, deriving from iron and manganese, varies in intensity, growing darker over time. In its newly excreted state the test is actually colorless.

What I most often see of *Arcella* is *A. vulgaris* (figures 2.47, 2.48, and 2.49). From the ventral side a recessed aperture is visible as well as the honeycomb surface of the test. A lateral view reveals the test's characteristic dome

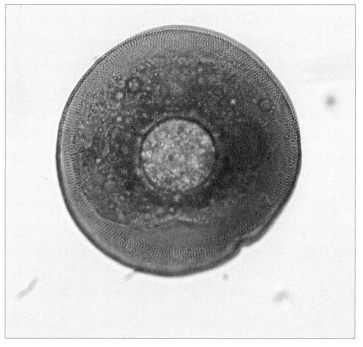

Figure 2.47. *Arcella vulgaris* showing the aperture and honeycomb surface of the test. ~ 100 μm, 400x

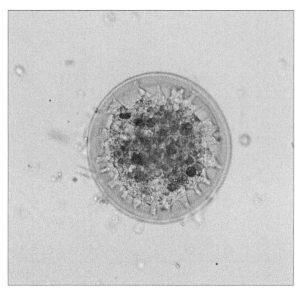

Figure 2.48. Aboral view of *Arcella vulgaris*. Note the small pseudopodia-like structures anchoring the cell to the sides of the test. ~ 90 μm, 200x

Figure 2.49. Lateral view of *Arcella vulgaris* showing the dome shape of the test. 200x

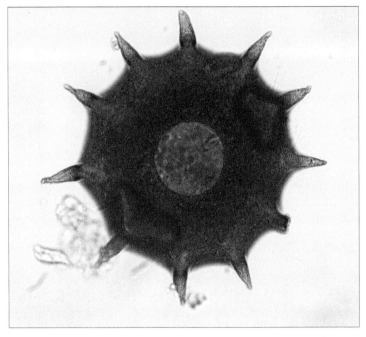

Figure 2.50. *Arcella dentata*, ventral view. ~ 117 µm, 400x

shape. Also common in my samples are lovely *A. dentata* (Figure 2.50) and *A. hemisphaerica* (Figure 2.51).

Nematodes too are present today and in rather large numbers. They are roundworms, non-segmented, and, in the case of freshwater species, usually less than a centimeter long (Figure 2.52). Their movement in a concavity-slide is a seemingly frantic side-to-side thrashing. It is, perhaps, less fervid under normal living conditions.

Nematodes inhabit the bottom mud, sand, and detritus of freshwater environments where they often constitute the greatest percentage of biomass. Their diet,

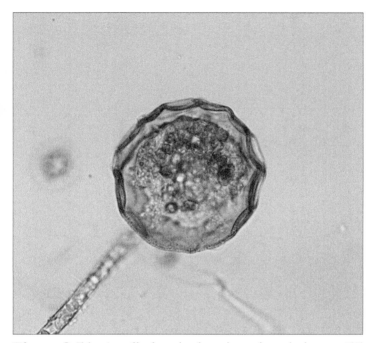

Figure 2.51. *Arcella hemisphaerica*, aboral view. ~ 55 μm, 400x

depending upon species, ranges from living plant material, to detritus, to each other.

As is usually the case, today I also see an organism that is new to me. In this instance it has a clear ovoid body with a sweeping anterior flagellum and a very long stationary "tail." I try to take photos but the results are too blurry to be useful. However, I make a very basic drawing of what I see with a written description in my notebook.

In addition to a journal, I keep an 8 x 11-inch notebook for all of my samples in which I record date, time, locality, water temperature and pH, weather conditions,

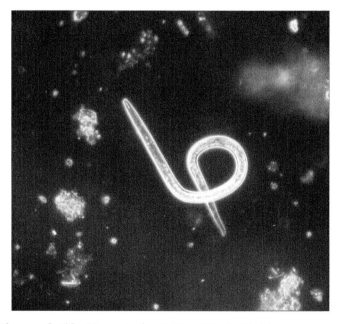

Figure 2.52. Nematode. 100x, Dark Field

measurements of specimens, and some drawings of what I observe. The drawings, crude as they are, have often helped with identification when photographs are poor or lacking. Trying to draw what I see has also improved my observation skills by forcing an appreciation for detail. Admittedly, keeping notebooks is a compulsion, but over the years their personal worth has more than repaid their bulk. It also indulges the simple, anachronistic pleasure of applying a pencil to a sheet of paper.

Thursday, September 13th, 3:00 P.M.

Manor Kill Falls (Figure 2.53). I take a drive to this pretty place that I have visited so many times in the past, not for water samples, but for Middle Devonian-age plant fossils. Today I am focused on water. The falls are running at a slower velocity so I am able to approach them without being drenched. Smoothed rocks at the base are covered with *Cladophora*, the filaments pressed over their surfaces in the direction of flow. They give the odd appearance of many detached human heads with wet green hair. I take a small sample of *Cladophora* and some water and wend my way homeward.

The sample is replete with pennate diatoms, small bdelloid rotifers grazing the algal filaments, and a background of very tiny round forms. But the jewel of this sample is *Vorticella* attached to *Cladophora*, its stalk fully extended and the bell of its body crystal clear against my dark field background (Figure 2.54). *Vorticella* is so

Figure 2.53. Manor Kill Falls.

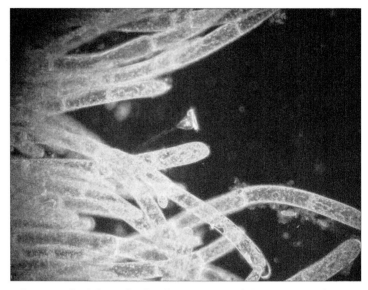

Figure 2.54. *Cladophora* sp. with *Vorticella* sp. attached. 100x, Dark Field

common — I have seen it hundreds of times — but the thrill of an encounter never diminishes. Especially today, when considering the force of water from the falls. How does something this small and delicate endure so perfectly? I may see *Vorticella* again a thousand times but that ever-increasing familiarity will not depreciate its singular beauty.

Autumn

Friday, October 3rd, 2:00 P.M.

Sikule Pond. This is a large nearby pond that is a seasonal home for swans, geese, ducks, blue herons, green herons, and the occasional bald eagle (Figure 3.1). I take a water sample from among the cattails but have to slog through the mud to get it as the water level is low and beyond my reach from the shore.

What catch my attention first under the microscope are the many colonies of *Coelosphaerium*. However, no

Figure 3.1. Sikule Pond. Swans are just visible in the right-hand portion of the photo.

other cyanobacteria seem to be present, nor is there any other evidence of a bloom as occurred at Lawson Lake in August. There are many diatoms, *Cymbella* the most common. *Cystodinium* is present and I find the common, but no less lovely, desmid *Cosmarium* (Figure 3.2).

My second slide holds a loricate rotifer that is new to me — a species of *Trichotria* (Figure 3.3). It is large at roughly 700 µm with two very long toes of equal length. Another loricate rotifer, frequently encountered in my samples, is also present — a species of *Lecane* (Figure 3.4).

Additionally, I observe *Arcella discoides*. It is not an empty test as is usually the case, but a test pretty well filled with its amoeba resident. However, the presence of four air bubbles within the test makes it appear decidedly odd (Figure 3.5).

The range of protist morphologies and habits is astounding and at times one forgets that they are generally single cells. Such is the case with *Lacrymaria olor*

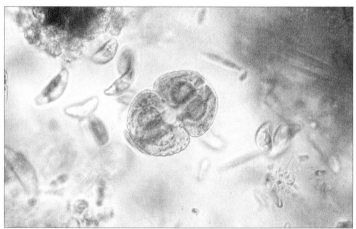

Figure 3.2. *Cosmarium* sp. ~100 µm, 200x

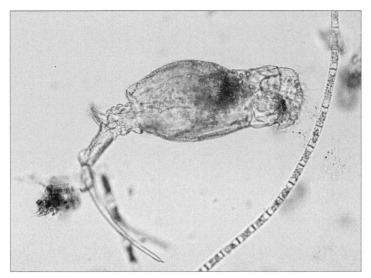

Figure 3.3. *Trichotria* sp., lateral view. ~ 700 µm, 100x

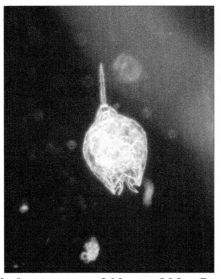

Figure 3.4. *Lecane* sp. ~ 260 µm, 200x, Dark Field

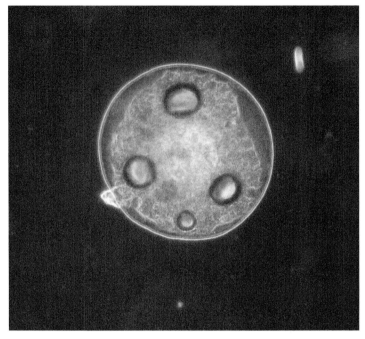

Figure 3.5. *Arcella discoides* with air bubbles. 200x, Dark Field

(Figure 3.6) which I find in my next slide. A ciliated protist, the cilia are arranged in rows spirally over the roughly 200 μm of the cell's length. However, it is the shape of the cell that is so striking; it appears to have a tapered body and a long neck. The "neck," bearing a mouth, can extend into a thin band many times the length of the "body." It assumes a serpentine quality as it searches for prey, winding easily around any object in its path, lunging and contracting at a frantic rate. *Lacrymaria* is one cell and yet its efficacy places it on a par with any species of predatory animal.

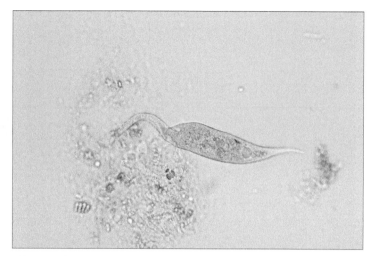

Figure 3.6. *Lacrymaria olor.* ~ 200 μm, 200x

Sunday, October 5th, 10:30 A.M.

Sikule Pond. Overcast and calm. Lots of geese, a heron, and a pair of swans congregate toward the far side of the pond.

My first encounter under the microscope is with the amoeboid protist *Actinosphaerium* (Figure 3.7). This is a large heliozoan distinguished by a striking peripheral layer of vacuoles which expel excess water from the cell. The radiating axopodia taper from their bases outward, and, as in *Acanthocystis*, passively trap prey. Prey for *Actinosphaerium* are larger, however, and include rotifers and copepods.

Testatae amoebae are also present and I find a nice example of *Difflugia acuminata* (Figure 3.8). This species of *Difflugia* is easily identified by a pointy little posterior extension of its gracefully shaped test. The test

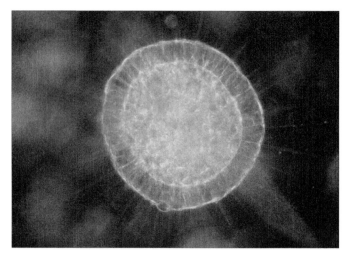

Figure 3.7. *Actinosphaerium* sp. Body ~ 325 µm, 100x, Dark Field

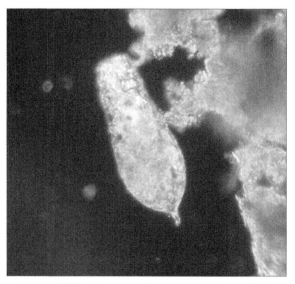

Figure 3.8. *Difflugia acuminata.* ~ 175 µm, 200x, Dark Field

itself is composed of accreted mineral particles, probably quartz, but, according to Joseph Leidy, *D. acuminata* tests may also comprise diatom frustules.

Normally when using a light microscope bacteria are not observed as they are generally colorless and beyond the resolution powers of standard objectives; stains and oil immersion objectives are required. However, nearly every sample from Sikule Pond provides me an opportunity to view two types of bacteria as one type is very deep pink in color and the other is extremely large.

The pink bacteria are *Lamprocystis* sp. (Figure 3.9). These photosynthetic cells are ovoid or spherical in shape and 2–3 µm in diameter. The pink color derives from

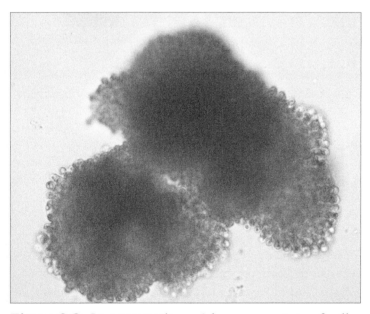

Figure 3.9. *Lamprocystis* sp. A large aggregate of cells. 100x

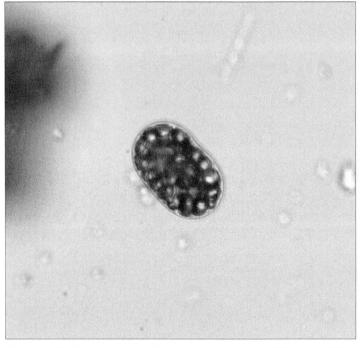

Figure 3.10. *Achromatium oxaliferum.* ~ 55 μm, 200x

carotenoid pigments and the cells form aggregates embedded in mucilage. These cell aggregates can be substantial in size and striking in their bright pink color which allows for their observation.

The unusually large bacteria are *Achromatium oxaliferum* (Figure 3.10). As the genus name indicates, these bacteria are without color yet they are noticeable because of their enormous size. They appear as single cells, round or oval in shape, with conspicuous intracellular calcium carbonate crystals. Achieving lengths of 50–100 μm, they are among the largest known bacteria.

Both *Lamprocystis* sp. and *Achromatium oxaliferum* are "sulfur bacteria" that utilize sulfide oxidation for energy production.

Saturday, October 26th, 2:30 P.M.

Sikule Pond. A cloudy, windy, chilly day. Dozens of geese are on the pond and the pair of swans is still here. I take a sample far to the right of my usual area as the birds are close and I am reluctant to disturb them. It is good to get back into the warm car, away from the wind, and I remember that October is nearly over. A month from today will be the Thanksgiving holiday; my good sampling time is slipping quickly away.

The first slide contains a familiar ciliate — *Stylonychia* (Figure 3.11). It is of moderate size and has two distinguishing features. One is the band of

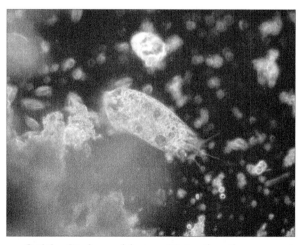

Figure 3.11. *Stylonychia* sp. Note the three posterior cirri. ~ 170 µm, 100x, Dark Field

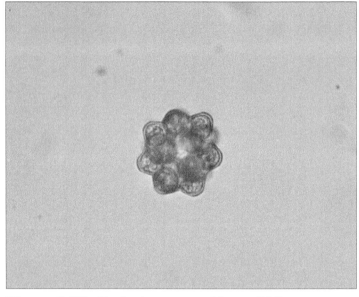

Figure 3.12. *Coelastrum* sp. ~ 40 μm, 200x

membranelles or specialized cilia on the left side of its oral aperture which gives the appearance of a collar. The other character is the presence of three posterior cirri or slender tufts that are very easily observed.

Two common non-motile green algae are also present in today's sample. The first is *Coelastrum* (Figure 3.12), a coenobial colony of spherical cells that are connected to each other via blunt projections of their surrounding mucilaginous sheaths. The colony is attractively star-shaped and dark green. The second alga is *Selenastrum* (Figure 3.13). This is a loose colony of crescent-shaped cells that are attached back-to-back. Both *Coelastrum* and *Selenastrum* are planktonic forms.

Familiarity gives way to novelty in my next slide. I am introduced to a peritrich ciliate that is more robust than the usual *Vorticella* I see, and its stalk lacks the ability to contract. This is probably a species of *Campanella* (Figure 3.14). Like all peritrich ciliates it is a wonder to observe, but it lacks the grace and charm of *Vorticella*.

My last slide is highlighted by *Dinobryon* (Figure 3.15), and *Gloeotrichia* (Figure 3.16). *Dinobryon* is a colonial, planktonic alga characterized by a branch-like colony. The protoplasm of each cell resides within a lorica made of cellulose microfibrils secreted by the cell. The lorica can be vase-shaped, funnel-shaped, or cylindrical depending on the species. The addition of a new lorica by a daughter cell just inside the anterior portion of the parent lorica continues the branch form of the colony.

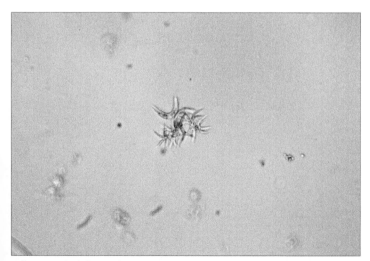

Figure 3.13. *Selenastrum* sp. 200x

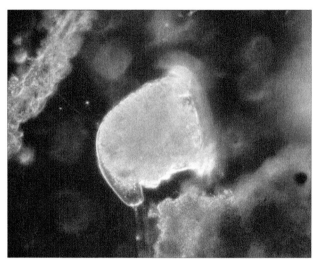

Figure 3.14. *Campanella* sp. with oral cilia "whirring." ~ 185 μm, 200x, Dark Field

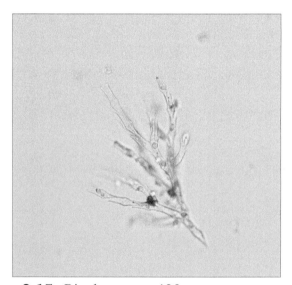

Figure 3.15. *Dinobryon* sp. 400x

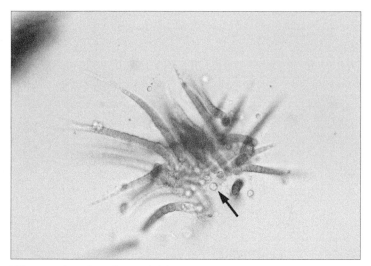

Figure 3.16. *Gloeotrichia* sp. 200x

In addition to basic cellular organelles, cells comprise two unequal flagella, a stigma or eyespot, and one or two chloroplasts. The latter components allow for a phototrophic habit, but *Dinobryon* cells are also phagotrophic, their prey a variety of bacteria.

Gloeotrichia is a colonial cyanobacterium that appears in my sample in a typical brown color. Round heterocysts (nitrogen-fixing cells) are clearly evident at the basal ends of some of the filaments.

This last slide also holds a testate amoeba that appears to be *Arcella conica* (Figure 3.17). The test is dome-shaped and light brown in color, with angular facets on its dorsal surface. There is no equivocating; it is simply an elegant little structure.

Additionally, I see an example of one of the largest of the ciliated protists, *Spirostomum* (Figure 3.18). At first

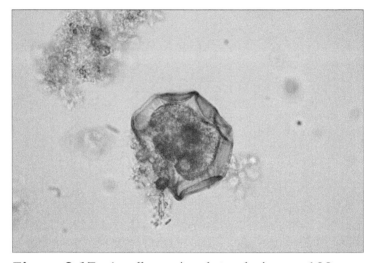

Figure 3.17. *Arcella conica*, lateral view. ~ 100 µm, 200x

glance this organism may appear to be a worm. However, close observation will reveal parallel rows of cilia along the entire length of the cell. *Spirostomum* is best known for its millisecond contraction capability when disturbed.

On my return to the pond with the sample the geese have all gone, perhaps to begin their autumnal journey southward. The swans remain, gliding forth and back in semi-circles, and I wonder when their date of departure will be. Micro-communities too will change with the coming cold. In the months ahead I will slowly note less diversity and fewer individuals. Some cells will form resistant cysts or adopt other strategies to cope with the cold and changes in light and oxygen levels. Likewise the tiny metazoans.

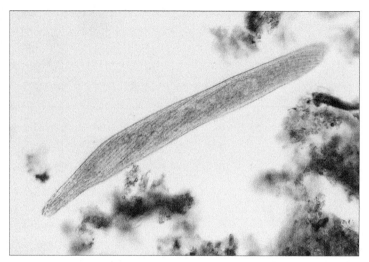

Figure 3.18. *Spirostomum* sp. Note the rows of cilia along the length of the cell. ~ 450 μm, 100x

Saturday, November 16ᵗʰ, 12:35 P.M.

Rugg Road Shrub Swamp. The water has turned to slush after a night of sharp cold but the lowered temperature has not yet really affected the diversity of the swamp. I see *Spirogyra, Nostoc, Cystodinium,* ostracods, copepods, and nematodes. A beautiful *Chlorohydra* dominates my second slide. And, as usual, I find an abundance of testate amoebae, particularly *Arcella,* although many of the tests appear to be empty. I also encounter two naked amoebae. The first one is larger than the second, but both exhibit cylindrical pseudopodia with rounded ends and cytoplasmic crystals. The cytoplasmic streaming that facilitates movement is evident in both specimens. They are beautiful organisms! Against the dark field illumination of my microscope they fairly

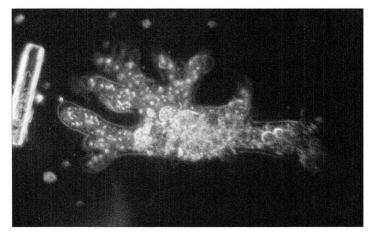

Figure 3.19. ?*Polychaos* sp. The amoeba has consumed several organisms. 200x, Dark Field

glitter with light reflected from the crystals. While I am not by any means certain, my specimens may be a species of *Polychaos* (Figure 3.19).

Even though it is mid-November this little swamp does not disappoint. In addition to the amoebae and the variety of other organisms observed today I also have the privilege of seeing *Stentor* (Figure 3.20). A large ciliate whose cell is uniformly covered with short cilia, *Stentor* assumes a distinctive trumpet shape when feeding. However, disturbance causes it to quickly contract into a protective mucous sheath at the cell's basal end. This region of the cell also provides a holdfast for attaching to substrates while feeding. At times, the cell may release its hold and swim to a better feeding area. Its form then becomes oval or pear-shaped (Figure 3.21), its swimming movement a rapid spin or roll.

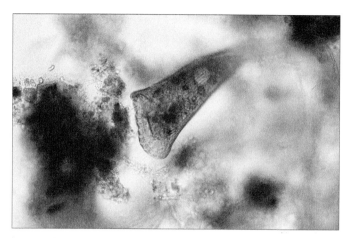

Figure 3.20. *Stentor* sp. in its "trumpet shape" while feeding. 100x

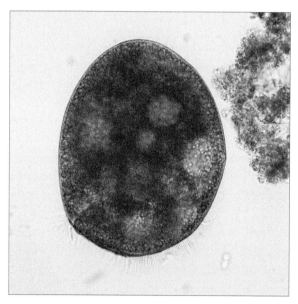

Figure 3.21. *Stentor* sp. The contracted swimming form of a very dark green individual. 200x

Stentor cells may be blue, brown, black, pink, green, or may lack color altogether. Today's specimen is bluish green, but more often the cells I encounter are a very light blue.

It is difficult to return my sample to the swamp; I hesitate several moments before pouring it back in. Many of the organisms I have observed today will not be seen again until spring.

Wednesday, November 27th, 10:30 A.M.

Rugg Road Shrub Swamp. I make a quick run for sample to take advantage of this rare, 63° day. A layer of thin ice outlines the water's edge from successive nights of heavy frost, but otherwise the surface is free.

Rather diminished activity in the water today with *Euplotes daidaleos* and *Paramecium bursaria* the dominant organisms. I am once again reminded of the beauty of *Paramecium bursaria* ciliates, their interiors verdant with *Chlorella*, their movements elegant and smooth as they glide within a thick background of plant debris.

As a large, ungainly shape suddenly begins to fill my field of view I am also reminded how extraordinarily diverse micro-communities can be. A pair of tiny black eyes within a mottled brown, undulating body identify this resident as a flatworm (turbellarian) (Figure 3.22). It is not a pretty thing, but it is splendid in its size and plasticity of form. It contracts, expands, and twists in a series of fluid movements, finally disappearing entirely within the background of debris. Although measuring nearly 900 μm, this individual is considered

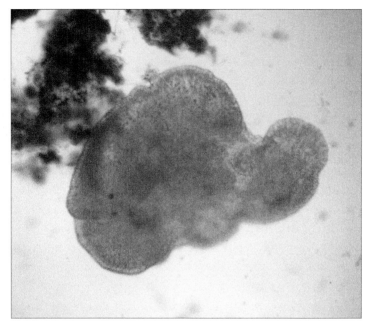

Figure 3.22. Turbellarian flatworm. 200x

a *micro*turbellarian as it is under one millimeter in length. *Macro*turbellarians are quite a bit larger and include the well-known planarians.

While observing my specimen's movements some fullness of body is evident; it is not flat. Indeed, many species of turbellaria are not supine regardless of their flatworm designation.

A characteristic that *is* common to all turbellaria is a ciliated epidermis. It is by means of cilia that the animals move about, microturbellarians feeding on bacteria, protists, and other small invertebrates while their macro counterparts prey on larger organisms.

Saturday, December 7ᵗʰ, 1:45 P.M.

Rugg Road Shrub Swamp. 30°. A partly sunny day after a very dreary night of sleet and snow. I find a quarter-inch layer of ice to chop through to obtain my sample; ice is floating in my aquarium still as I prepare my first slide.

Very diminished activity and diversity. I see a spiny, green *Cosmarium* and a lunate *Closterium*. Also, a few pinnularid diatoms. There are many amoeba tests, mostly *Arcella*, but all are empty. Slide after slide exhibits nearly the same population of organisms and I am about to call it a day when I spot a tardigrade. It is not a live tardigrade, however. What I have chanced upon is the shed cuticle of a tardigrade — an exuvium — containing seven large, mature eggs (Figure 3.23).

Tardigrades are very small (50 μm to 1200 μm) invertebrate animals taxonomically classified under the phylum Tardigrada (Figure 3.24). This placement within a separate phylum is the result of much historical uncertainty regarding their affinity. Since discovery in 1773, tardigrades have been at various times associated with nematodes, crustaceans, velvet worms, rotifers, and other groups. However, recent molecular studies have shown them to be a sister group of arthropods.

Aquatic animals, tardigrades are found in both freshwater and marine environments as well as moist terrestrial habitats such as those found within mosses. The vernacular names "water bear" and "moss piglet" refer to body shape, the most distinguishing feature being four pairs of short, stubby legs with claws. The name "tardigrade" means slow walker and, in fact, their movement is usually restricted to a lumbering walk; they do not swim.

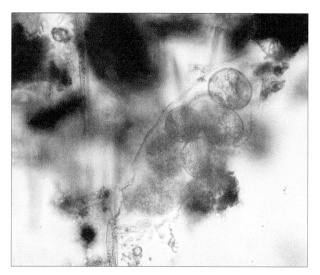

Figure 3.23. Shed cuticle (exuvium) of a tardigrade with seven eggs. 100x

Figure 3.24. Tardigrade, photographed on an earlier date from The Swamp. ~ 900 μm, 40x

While most species of freshwater tardigrades are white or colorless, terrestrial species may exhibit a range of colors due to pigmentation or as a consequence of food sources. Food items include algae, bryophytes, protozoa, rotifers, nematodes, and smaller tardigrades, all from which they suck fluids.

The claws of tardigrades are an important morphological character used in determining species. During the molts that occur throughout the animal's life the entire cuticle is shed, including the claws. It is common for egg-laying to be included in a molt as I have encountered in my sample today. However, eggs are also often laid freely in the substratum.

Since all tardigrades are aquatic, they are subject to changing environmental conditions such as drying and temperature extremes. To cope with these issues they have developed the ability to withstand periods of stress by entering one of five latent states during which metabolism and other life functions are severely reduced or suspended. One latent state, anhydrobiosis, causes the animal to shrink into a barrel-shaped form called a tun. Tuns are able to defy the effects of extreme heat, cold, pressure, radiation, and toxic chemicals. They have even withstood the extraordinary rigors of space travel. Regardless of the type of stress, rehydration of a tun will have a tardigrade resuming normal function in a matter of minutes, some reputedly having been restored to life after intervals of several years. This amazing phenomenon is due, in large measure, to tardigrade-specific proteins that are augmented during desiccation and function to increase the animal's desiccation tolerance.

Truly, the roly-poly tardigrades are marvelous survival strategists. But their success begs the question: Why have not all organisms developed comparable strategies? Some have — the aforementioned rotifers, for example. But only tardigrades have evolved such a broad *suite* of strategies, the means to withstand just about anything. While the fossil record for these animals is very poor, it is believed that they arose in the Cambrian Period, some 500 million years ago.

This extremely long history has given them a leg up, so to speak. That is, they have had plenty of time to "tinker."

Tardigrades have achieved the apex in survival planning without the need for "how to" manuals and bunkers. When and if the dreaded Apocalypse does occur, a tiny tardigrade may be the last "man" standing.

Winter

Monday, January 25*th*, 10:00 A.M.

Rugg Road Shrub Swamp. It is 24° and a nor'easter looms, the sky already becoming a solid white dome. I hurriedly chop through 4 inches of ice with a hammer and chisel and finally reach a few inches of water. The beaker end of my sample collector just fits the diameter of the chopped hole and I bring up enough water to half-fill my container. It would, perhaps, have been wiser to stay within doors with a good book, but I am curious to see who might be out and about in the water on this day, the epitome of winter in the Northeast.

The first slide — very, very sparse. I see a small rotifer of some kind and two *Peridinium* dinoflagellates moving slowly. Slide two is even more spare; besides a background of extremely tiny twirling organisms, all I encounter is a pennate diatom. However, the diatom is beautifully golden with an abundance of oil droplets inside. Slide three holds a nematode thrashing only slightly less frantically than normal, and *Chaetonotus* racing by as if pursued by a wee white shark. Succeeding slides show little more, but the very last holds . . . *Vorticella*. Anchored to a mat of plant debris with stalk extended, its cilia beat at the same tempo as if it were a sunny summer day. And, as usual, I am delighted to see it. But today, seeing it thrive regardless of the extreme weather, is more about comfort. With change occurring globally at an unsettling rate my psyche needs all the constancy it can get. I can depend on *Vorticella*. No matter where, no matter when, this amazing cell beats on.

I return to the swamp. Snow is falling steadily now and the Catskill Mountains, usually prominent in the south, have vanished from view. I re-chop the hole in the ice and pour the water back in. The silence of the big storm returns. Tomorrow the swamp too will be lost in snow, the water completely beyond my reach.

Tuesday, February 23rd, 1:35 P.M.

Ostracod Ditch. 15°. Temperatures the past three nights have dipped to 6° below zero — a late-winter cold snap. I chop through the ice near a culvert in the hope of finding a bit of water (Figure 4.1). As I chop, a passing truck slows and stops. It is a member of the town's road crew. He kindly inquires whether I am all right and I assure him that I am. There is an awkward pause. I know he wonders why I am on my hands and knees on the ice in the ditch but he is, perhaps, too embarrassed to ask. For my part, I am too tongue-tied to explain. I smile, he smiles. We wave and off he goes. I retrieve a small sample and hurry home.

My first slide reveals a predominance of green algal filaments. The sight of living green is a jolt after being surrounded by a white world for the past three months. *Vaucheria*, *Spirogyra*, and a very thin filament, unknown to me, are present in every slide. Although it is still winter, life is on the threshold of renewal. Next month the vernal sun will begin to warm the waters. Micro-communities will flourish and continue their role in helping to

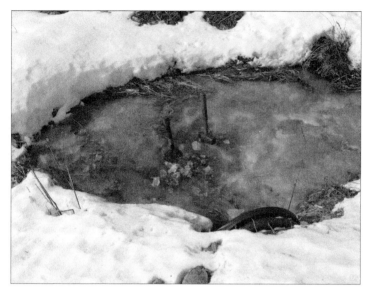

Figure 4.1. Frozen Ostracod Ditch.

maintain aquatic habitats. By producing oxygen, purify-
ing water through consumption of bacteria and detritus,
being part of the food chain, and contributing to the carbon cycle, these communities are *vital* to habitat stability.

Warmer weather will also induce me to continue
my rounds, sampling all the local water sources. As a
consequence of years of this activity I have come to believe that, while water environments everywhere must
unfailingly be conserved and protected, they should also
be *revered*. Many of the microorganisms we see in them
today, often stunning in their intricacy, variety, and
beauty, are, more importantly, the runes and glyphs of
some of life's most ancient history. Water is, after all, our
true ancestral home.

Glossary

Aboral – An aspect of an organism that is farthest away from the position of the mouth.

Agar plate – A Petri dish containing a growth medium for culturing bacteria, fungi, etc.

Amoeba – A protist lacking a specific form and using pseudopodia for locomotion and feeding. (plural = amoebae)

Amoeboid – Resembling an amoeba.

Anhydrobiosis — A dormant state in which an organism becomes dehydrated, thus greatly reducing its metabolic activity.

Anterior – The front of an organism (as opposed to the posterior).

Antheridium – The male reproductive structure in certain plants and algae. (plural = antheridia)

Autotroph – An organism that synthesizes its food from inorganic sources through the use of light or chemical energy.

Basal – Referring to the base or point of attachment of an organism.

Benthos – The community of organisms living on the bottom of a given water body.

Biomass – The total mass of living matter comprising a specific habitat.

Carapace – The hard, outer covering of certain invertebrate animals.

Carotenoid pigment – A yellow or orange pigment that usually acts as an accessory pigment in photosynthesis.

Cauda – A tail-like extension on the posterior portion of an organism. (plural = caudae)

Centric – In diatoms, those frustules that are radially symmetrical.

Chloroplast – A plastid (cellular organelle involved in food-making) that contains chlorophyll and is the site of photosynthesis.

Ciliate – A protist characterized by the presence of many short cilia on its outer surface.

Cilium – A bristle-like structure that protrudes from the surface of a cell to help facilitate movement. (plural = cilia)

Cirrus – A structure that facilitates locomotion and is characteristic of some ciliate protists. A cirrus is formed from a compact cluster of cilia. (plural = cirri)

Coenobium – An algal colony that contains a fixed number of cells. (plural = coenobia)

Concavity slide – A rather thick glass microscope slide having a depression for holding sample. Also known as a "well slide" or "culture slide."

Cortex – The outer covering of a cell.

Cosmopolitan distribution – When the range of an organism, associated with its particular habitat, extends across most of the world.

Cyanobacterium – A photosynthetic bacterium having

a blue-green color, formerly (and erroneously) known as a blue-green alga. (plural = cyanobacteria)

Cytoplasm – The material inside a living cell within the cell membrane. It contains the cellular organelles such as mitochondria, etc., and surrounds the nucleus. (see Protoplasm)

Desmid – An algal protist consisting of two semi-cells sharing a nucleus.

Detritus – Organic waste material produced by the decomposition of dead organisms.

Diatom – An algal protist having an outer covering or frustule made of silica.

Dinoflagellate – A planktonic protist having two flagella.

Dorsal – Refers to the upper side of an organism.

Endosymbiont – An organism that lives inside another organism under symbiotic or mutually beneficial conditions.

Epibiont – An organism that lives on the surface of another organism.

Epilithic – Living or growing on the surface of rocks.

Euglenoid – A flagellated protist included in the phylum Euglenozoa. It may be green and photosynthetic or colorless and non-photosynthetic.

Eukaryote – An organism that has a membrane-bounded nucleus and membrane-bounded organelles. Eukaryotes comprise all organisms except bacteria.

Eutrophic – Pertains to a body of water rich in nutrients which spurs an increased rate of growth in its plants, algae, and other organisms.

Flagellate – A protist bearing one or more flagella.

Flagellum – A whip-like extension of some cells that facilitates locomotion. (plural = flagella)

Frustule – The outer covering of a diatom, composed of two overlapping valves made of silica.

Fusiform – Spindle-shaped.

Gastrotrich – A small invertebrate animal, common in freshwater, having bands of cilia on its ventral surface and two posterior projections that contain adhesive glands.

Genus – The taxonomic rank immediately above "species." (plural = genera)

Girdle – In diatoms, the space between or meeting point between the two frustule valves.

Gonidium – In algae, an asexual cell or group of cells. (plural = gonidia)

Heliozoan – An amoeboid protist with a spherical cell and radiating axopodia.

Heterotroph – An organism that cannot manufacture its own food supply; it relies on the consumption of other organisms or other forms of organic material to survive.

Hyaloplasm – The outermost, clear, and rather rigid cytoplasm of an amoeba pseudopod.

Light microscope – The most common form of microscope. It uses focused light and a series of magnifying lenses to enlarge tiny forms for observation.

Liverwort – A small, non-vascular plant usually found in moist habitats often in the company of mosses.

While habitat and reproduction are similar in liverworts and mosses the plants differ in several significant ways.

Lorica – The protective outer sheath of rotifers and some protists.

Mastax – The esophageal area of a rotifer.

Membranelle – In ciliate protists, the specialized cilia associated with the oral aperture.

Metaboly – The ability of a euglenoid protist cell to change its shape by means of peristaltic or wave-like movements.

Metazoan – A eukaryotic organism (animal) having a multicellular body with cells differentiated into tissues and organs. (plural = metazoa)

Microfibril – A tiny, thread-like, proteinaceous structure common in eukaryotic cells.

Micrometer – A unit of measure equal to one millionth of a meter and denoted by the symbol "μm." Also known as a "micron."

Millimeter – A unit of measure equal to one thousandth of a meter and denoted by the symbol "mm."

Milliliter – A unit of measure equal to one thousandth of a liter and denoted by the symbol "ml."

Mitotic – Refers to mitosis which is a form of cell division in which two daughter cells are produced having the same number of chromosomes as the parent.

Monopodial – In reference to amoebae, when an amoeba lacks pseudopodia and the entire cell acts as a single pseudopodium.

Morphology – The form of a living organism.

Motile – Capable of motion.

Oogonium – In algae, the female sex organ that contains one or more eggs. (plural = oogonia)

Organelle – One of several specialized structures within a living cell.

Paramylon – The nutritional reserve of euglenoids in the form of glucose. Paramylon may be observed within euglenoid cells as "paramylon bodies" of various shapes.

Parthenogenesis – A form of asexual reproduction in which the egg is not fertilized.

Pellicle – The outer surface of euglenoids and some other protists.

Pennate – In diatoms, those frustules that are bilaterally symmetrical.

Peritrich – A ciliate protist, usually bell-shaped, bearing an oral ring of cilia.

Phagotrophic – Feeding by engulfing a food item and ingesting it by means of a phagocytic vacuole.

Phototrophic – The use of light to synthesize food from inorganic sources, as in photosynthesis.

Phylum – A taxonomic rank just below a kingdom.

Planarian – A type of flatworm (Turbellarian) usually measuring around 3–15 mm although some may be even larger.

Planktonic – Refers to free-floating microscopic organisms in freshwater and marine environments.

Plastid – The photosynthetic organelle in protists and plants.

Polymer – A molecular structure having chains of many similar units.

Protoplasm – The material within a living cell that includes the cell membrane, cytoplasm, and nucleus. Cytoplasm is a *part* of the protoplasm.

Pseudopodium – A temporary extension of the cytoplasm in an amoeba for the purposes of locomotion and feeding. (plural = pseudopodia)

Pyrenoid – A proteinaccous structure associated with a chloroplast that is involved in carbon fixation, starch formation, and storage.

Sessile – Immobile, fixed in one place.

Seta – A stiff, bristle-like structure on the outer covering of an organism. (plural = setae)

Siliceous – Made of silica.

Somatic cell – Any body cell of an organism with the exception of a reproductive cell.

Test – A hard, outer covering of an organism. A shell. Amoebae having an outer shell are described as being "testate" as opposed to "naked" in which a shell is lacking.

Thalloid – In liverworts, refers to species having a flat aspect of growth from a thallus as opposed to leafy species.

Trophi – A set of hard jaws located in the mastax or esophageal area of rotifers.

Tussock sedge (*Carex stricta*) – A species of sedge that grows in raised clumps in swamps, marshes, and other wetlands.

Uroid – The posterior protuberance of a moving amoeba.

Vacuole – An open space within the cytoplasm of a cell that contains fluid or air and is surrounded by a membrane. Vacuoles serve a number of different functions depending on the type of cell.

Valve – The top or bottom component of a diatom frustule.

Ventral – Refers to the underside of an organism.

Zygote – A cell that is formed from the fusion of an egg cell and a sperm cell.

Bibliography

AlgaeBase. World-wide electronic publication, National University of Ireland, Galway.
   http://www.algaebase.org

Bailey, J. W. 1843. Infusoria. Pp.48–79 in William W. Mather, *Geology of New York, Part I, Comprising the First District*. Albany: Carroll & Cook, Printers to the Assembly.

Bailey, J. W. 1850. Microscopical observations made in South Carolina, Georgia, and Florida. *Smithsonian Contributions to Knowledge*, Volume II.

Bailey, J. W. 1855. Reply to some remarks by W. H. Wenham, and notice of a new locality of microscopic test-object. *The American Journal of Science and Arts*, Second Series, Volume XIX, 28–30.

Bellinger, E. G., and D. C. Sigee. 2010. *Freshwater Algae: Identification and Use as Bioindicators*. Hoboken: John Wiley & Sons.

Boardman, R. S., A. H. Cheetham, and A. J. Rowell, eds. 1987. *Fossil Invertebrates*. Boston: Blackwell Scientific Publications.

Boo, Sung Min, et al. 2010. Complex phylogenetic patterns in the freshwater alga *Synura* provide new insights into ubiquity vs. endemism in microbial eukaryotes. *Molecular Ecology* 19:4328–4338.

Boothby, T. C., et al. 2017. Tardigrades use intrinsically disordered proteins to survive desiccation. *Molecular Cell* 65:975–984.

Brook, A. J. 1980. Barium accumulation by desmids of the genus *Closterium (Zygnemaphyceae). British Phycological Journal* 15:261–264.

Ciugulea, I., and R. E. Triemer. 2010. *A Color Atlas of Photosynthetic Euglenoids*. East Lansing: Michigan State University Press.

Coesel, P. F. M., and L. Krienitz. 2009. Diversity and geographic distribution of desmids and other coccoid green algae. Pp.147–158 in W. Foissner and D. L. Hawksworth, eds. *Protist Diversity and Geographical Distribution*. Heidelberg: Springer-Verlag.

Dillard, G. E. 1999. *Common Freshwater Algae of the United States*. Berlin: J. Cramer.

Diller, W. F., and D. Kounaris. 1966. Description of a zoochlorella-bearing form of *Euplotes, E. daidaleos* n. sp. (Ciliophora, Hypotrichida). *Biological Bulletin* 131(3):437–445.

Edgar, R. K. 1979. Jacob W. Bailey and the diatoms of the Wilkes Exploring Expedition (1838–1842). *Occasional Papers of the Farlow Herbarium of Cryptogamic Botany* 14:9–33.

Ehrenberg, C. G. 1843. *Verbreitung und Einfluss des Mikroskopischen Lebens in Süd – und Nord Amerika.* Berlin: Printworks of the Royal Academy of Sciences.

Foissner, W. 2009. Protist diversity and distribution: some basic considerations. Pp. 1–7 in W. Foissner and D. L. Hawksworth, eds. *Protist Diversity and*

Geographical Distribution. Heidelberg: Springer-Verlag.

Kinchin, I. M. 1994. *The Biology of Tardigrades.* London: Portland Press.

Kirk, D. L. 1998. *Volvox.* New York: Cambridge University Press.

Kreutz, M., and W. Foissner. 2006. The sphagnum ponds of Simmelried in Germany: A biodiversity hot-spot for microscopic organisms. *Protozoological Monographs,* Volume 3.

Lee, J. J., G. F. Leedale, and P. Bradbury, eds. 2000. *The Illustrated Guide to the Protozoa,* 2nd edition, Volumes I & II. Society of Protozoologists. Lawrence: Allen Press.

Leidy, J. 1879. *Freshwater Rhizopods of North America.* Washington, DC: Government Printing Office.

Patterson, D. J., and S. Hedley. 1992. *Free-Living Freshwater Protozoa.* London: Wolfe Publishing Ltd.

Penard Labs. *The Fascinating World of Amoebae.* Worldwide electronic publication, http://www.penard.de.

Pennak, R. W. 1989. *Freshwater Invertebrates of the United States,* 3rd edition. New York: John Wiley & Sons.

Raven, P. H., R. F. Evert, and S. E. Eichhorn. 1999. *Biology of Plants.* New York: W. H. Freeman and Company.

Ricci, C., and D. Fontaneto. 2009. The importance of being a bdelloid: ecological and evolutionary consequences of dormancy. *Italian Journal of Zoology* 76(3):240–249.

Ruiz, F., et al. 2013. Freshwater ostracods as environmental tracers. *International Journal of Environmental Science and Technology* 10:1115–1128.

Siemensma, F. J. 2015. *Microworld, the world of amoeboid organisms*. World-wide electronic publication, Kortenhoef, the Netherlands. http://www.arcella.nl

Stevens, D. C. 2013. The microscopic algae and water quality of the freshwater lake — the Little Sea, Studland, Dorset. *Quekett Journal of Microscopy* 42(1):45–68.

Sullivan, Navin. 1962. *Pioneer Germ Fighters*. New York: Scholastic Book Services.

Thorp, J. H., and A. P. Covich, eds. 2001. *Ecology and Classification of North American Freshwater Invertebrates*, 2nd edition. San Diego: Academic Press.

Van Aller Hernick, L., E. Landing, and K. E. Bartowski. 2008. Earth's oldest liverworts — *Metzgeriothallus sharonae* sp. nov. from the Middle Devonian (Givetian) of eastern New York. *Review of Palaeobotany and Palynology* 148:154–162.

van der Gast, Christopher J. 2014. Microbial biogeography: the end of the ubiquitous dispersal hypothesis? *Environmental Microbiology* 17: 544–546.

Vinyard, W. C. 1979. *Diatoms of North America*. Eureka, California: Mad River Press, Inc.

Warner, B. G., and R. Chengalath. 1988. Holocene fossil *Habrotrocha angusticollis* (Bdelloidea: Rotifera) in North America. *Journal of Paleolimnology* 1:141–147.

Warren, L. 1998. *Joseph Leidy: The Last Man Who Knew Everything*. New Haven & London: Yale University Press.

Wehr, J. D., R. G. Sheath, and J. P. Kociolek, eds. 2015. *Freshwater Algae of North America: Ecology and Classification*, 2nd edition. San Diego: Academic Press.

Whitman, W. 1892. *Prose Works*. Philadelphia: David McKay.

Wright, A. G., and D. H. Lynn. 1997. Maximum ages of ciliate lineages estimated using a small subunit rRNA molecular clock: crown eukaryotes date back to the Paleoproterozoic. *Archiv für Protistenkunde* 148:329–341.

Index

Italic page numbers denote illustrations

Epistylis, 73, *75*
Erie, Lake, 88, 90
Eudorina, 59, *61*
Euglenoid, 73, 75–78,
Euplotes daidaleos, 73, 118, *74*

Fingernail clams
(*Musculium securis*), 70
Flatworm, 118, *119*
Fragilaria, 39, 42

Gastrotrich, 67, *69*
Gloeotrichia, 111, 113, *113*
Gomphonema, 57–58, *57*
Gymnodinium, 21

Hyalotheca mucosa, 18, *19*

Lacrymaria olor, 102, 104, *105*
Lamprocystis, 107–109, *107*
Lawson Lake, 87, *88*
Lecane, 102, *103*
Leidy, Joseph, 49–52, 63–64, 107
Lepocinclis fusca, 76, *76*
acus, 92,
helicoideus, 92, *93*
Liverwort, 15–16

Manor Kill Falls, 98–99, *99*
Melosira, 67, *69*
Meridion, 35, 38, *37*

Micrasterias, 83, 85–86, *84, 85*
Microcystis, 88, 90
Mougeotia, 7, 38, 40, *41*

Naked amoeba, 52, 63, 66–67, *66*
Nauplius, 31–33, *32*
Nematode, 96–97, *98*
Nordlund Pond, 38, 57, *39*
Nostoc, 25, 27, *26, 27*

Ostracod, 7–11, 81, *9, 82*
Ostracod Ditch, 11, 63, 80, 126, *127*

Pandorina, 86, *86*
Paramecium bursaria, 80, 118, *81*
Parthenogenesis, 10
Patrick, Ruth, 69–70
Peridinium, 19–21, 125, *20*
pH, 30–31
Phacus longicauda, 77, *77*
gigas, 92, *93*
tortus, 92
Pine Pollen, 39, *43*
Pinnularia, *23*
Platyias patulus, 79, *79*
quadricornis, 79, *80*
Pleurotaenium, 60, *61*
Polychaos, 115–116, *116*
Protist, 16

Riccia fluitans, 15, 70, *15*